Iheb ESSOUSSI

Plant cultivation in forest nurseries & Reforestation

Iheb ESSOUSSI

Plant cultivation in forest nurseries & Reforestation

ScienciaScripts

Imprint
Any brand names and product names mentioned in this book are subject to trademark, brand or patent protection and are trademarks or registered trademarks of their respective holders. The use of brand names, product names, common names, trade names, product descriptions etc. even without a particular marking in this work is in no way to be construed to mean that such names may be regarded as unrestricted in respect of trademark and brand protection legislation and could thus be used by anyone.

Cover image: www.ingimage.com

This book is a translation from the original published under ISBN 978-620-2-54935-6.

Publisher:
Sciencia Scripts
is a trademark of
International Book Market Service Ltd., member of OmniScriptum Publishing Group
17 Meldrum Street, Beau Bassin 71504, Mauritius
Printed at: see last page
ISBN: 978-620-3-25718-2

Copyright © Iheb ESSOUSSI
Copyright © 2021 International Book Market Service Ltd., member of OmniScriptum Publishing Group

Culture des plants en pépinières

Chapter 1 :

Forest nurseries

Table of Contents

List of figures

1- Introduction :

The conservation of forest heritage is a vital requirement for maintaining ecological balance and preserving the forest heritage. To this end, policies have been introduced aimed at conserving and developing the various national forestry potentialities. Among these strategies is the installation of modern nurseries which enable the production of seedlings for reforestation which will later form the future forest.

The seedlings from nurseries are much more adapted and survive than seeds sown directly in place or by natural regeneration, due to the quality of the plant material produced through all stages from sowing to transplanting to the reforestation site.

2- The nursery :

2.1- Definition :

The nursery is a term that derives from the word "nursery", designating the land, the surface, the chosen and valued area devoted to the multiplication and breeding of plants until they can be planted elsewhere.

According to Ouédraogo (1989), the nursery is a specially selected and developed area where fruit trees or other fragile trees are sown and cared for.

The term nursery also refers to a company specialized in the production of woody, semi-woody and perennial plants. There are two main types of producers:

i-   **The producers of young seedlings:** who ensure the multiplication of plants by vegetative way (cutting, layering, in vitro multiplication...) or by sexual way (sowing). In general, they keep their subjects for one to two years at most.

ii- **The nurserymen-breeders:** who receive plants from the producers of young plants to then put them in culture over a minimum period of three years (except for perennials which have a cycle of less than one year). These plants are then sold directly to private individuals, communities, other nurseries or to companies that develop green spaces or possibly intended for the project of rehabilitation of deforested areas (Reforestation).

2.2- Type of nurseries :

There are two types of nurseries :

2.2.1- Temporary nurseries :

Which are located on the plantation site itself or in its vicinity. When the plants intended for planting have reached the desired size, the nursery is integrated into the planted site. This type of nursery is sometimes referred to as a **"flying nursery"**.

2.2.2- Permanent nurseries :

Which can be large or small depending on the objective and the number of seedlings grown each year. Small nurseries contain less than 100,000 seedlings at a time, while large nurseries contain more. In all cases, permanent nurseries must be well designed, located in a suitable site with an adequate water supply.

2.3- Nursery categories :

- **Category "A":** Nursery of prime importance, whose short-term rehabilitation is mandatory ;
- **Category "B":** Nursery of second importance that requires rehabilitation in the medium term ;
- **Category "C":** Nursery with major constraints and high management costs, which will be closed as the modernization of the nurseries of the two previous categories progresses.

2.4- Advantages of the nursery :

✓ Produce favorable conditions for seed germination and seedling development in order to have enough healthy seedlings.
✓ Provides optimal conditions for germination of seeds for the growth of valuable plants (temperature, light, air humidity, humidity of the substrate ...).
✓ Allows the selection of homogeneous plants before planting or sale.
✓ Facilitates the control and monitoring of young seedlings and their treatment against possible parasitic attacks.
✓ The nursery allows the production of off-season plants. Thus a saving of time because one does not need to wait for external conditions (Temperature and Precipitation) are adequate to the sowing.

2.5- Objectives of the nursery :

➤ **Short-term objective: To** have disease-free plant material suitable for reforestation. Consequently, nursery stock is much more adapted and survives much better than seeds sown directly in place or by natural regeneration.
➤ **Long-term objective:** The nursery provides quality seedlings for reforestation in order to restore and rehabilitate degraded forest ecosystems.

3- Nursery size :

The surface of the nursery must be sufficient to install the infrastructure, work, storage and drying areas. The nursery contains sections and each section formed by boards. The board is the smallest unit in the nursery.

The dimension of each board is generally one meter wide and 6 to 10 meters long. The boards can be sunk 30 to 35 centimetres below the general ground level. In this case, they can be lined with cement, stones or bricks. These boards contain the containers or have raised the plants.

To estimate total nursery area (TNA), the seedbed area is multiplied by 2.5 to account for access roads and service areas and 100 m² (for driveways) is added, based on a production of 2,000 plants per square metre of seedbed (**Figure 3).**

The size of the nursery and the total area of the nursery varies according to the diameter of the containers.

$$STP = (2.5 \text{ x seedbed area}) + 100 \text{ m}^2$$

❖ **Container diameter of 5 cm :**

For containers with a diameter of 5 centimeters, 240 square meters of planks are needed.

STP = (2.5 x 240) + 100 m² = 700 m²

❖ **Container diameter of 14 cm :**

STP = (2,5 x 1900) + 100 m² = 4850 ≈ 5000 m²

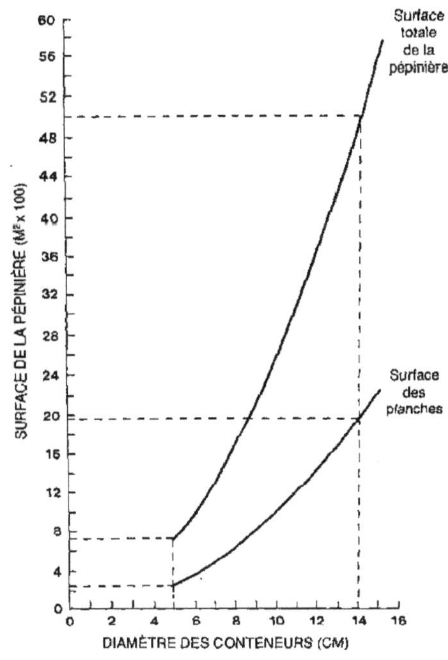

Figure 1: Estimate the total area of the nursery

4- Choice of site :

Before setting up a nursery, you must first choose an ideal location to obtain quality plants while minimizing costs. The choice of site for the installation of a nursery must meet several criteria.

4.1-   Accessibility to the road :

Accessibility is an important factor in the choice of setting up a forest nursery. Accessibility to the roadway facilitates the transportation of the plants from the nursery to the road, thus reducing the risks of damaging the plants during transport to the reforestation site.

4.2-   Proximity of the forest :

Another basic factor to consider when choosing a site is the availability of substrate (compost). Proximity to the reforestation site will reduce the time of exposure of the plants and their root systems to drying out.

4.3-   Proximity to water point :

The availability of water and its abundance in the nursery around the hour is unquestionable, especially during hot weather. Water is a substantial element in the breeding of the plants in watering and for the addition of additives in the water such as fertilizers for example. In this case we speak of **"fertigation"**, the combination of fertilization and irrigation.

Two aspects are important, water quality and daily water needs.

- **Water quality:** It must be slightly acidic, with a pH below 7, less than 550 parts/million dissolved salts and a conductivity below 0.8 mho/cm. In general, fairly soft and clear.
- **Quantity of water:** A sufficient quantity of water of the quality indicated above must be supplied daily to the nursery.

4.4-   Climate :

To have an optimal productivity of the plants and normalized, it is necessary to create a lenient condition for the young seedlings.

- Temperature should be between **20 and 25°C** ;
- Well exposed to sunlight ;
- Average humidity, well distributed throughout the year, to promote growth without hindering the completion of major works;
- Moderate wind that does not require special investments (reinforced staking, wind breaker...).

4.5-   Plot of land :

The ground must be friable, well-drained, flat or slightly sloping, with sandy and not stony soil that provides good stabilization for machinery and the necessary infrastructure. The soil must be as clean as possible (free of parasites and possible fungi that would damage the seedlings and seedling yolks).

4.6- Wind Breeze :

4.6.1- Definition :

Windbreaks (quickset hedges) are rows of trees and shrubs planted all around the nursery that serve to protect the young plants from the force of the wind, which is 4 to 5 metres high. Windbreaks should be kept at a distance so as not to shade (block) the crops. Windbreaks are usually made of **cypress, acacia or poplar.**

4.6.2- The benefits of windbreaks :
- ✓ Can reduce soil erosion,
- ✓ Increase agricultural production and protect livestock from heat and cold.
- ✓ Can house the buildings and corridors inside the nursery.
- ✓ Sources of wood and food ;
- ✓ Protects plants from the sun's burning rays
- ✓ Prevent livestock from entering the interior.

4.6.3- Function of the windbreak :
- ✓ Moderate soil and air temperatures ;
- ✓ Higher relative humidity ;
- ✓ Reduced evaporation and increased soil moisture.

4.6.4- Impacts of windbreaks on crop quality :
- ➢ Lowering the temperature during the day and raising it during the night;
- ➢ Increasing relative humidity and promoting soil moisture conservation ;
- ➢ Reducing damage caused by high winds ;

4.6.5- Impacts of windbreaks on the climate :
- ➢ Reduce climatic variations.
- ➢ Produce a more temperate microclimate on the scale of the protected plot; especially in winter and summer.
- ➢ Slowing down evapotranspiration, loss of water through the soil and the plant ;
- ➢ Maintain air humidity ;
- ➢ Slowing of the wind speed. The protection provided by a semi-permeable windbreak covers a distance of 10, 15 to 20 times its height.

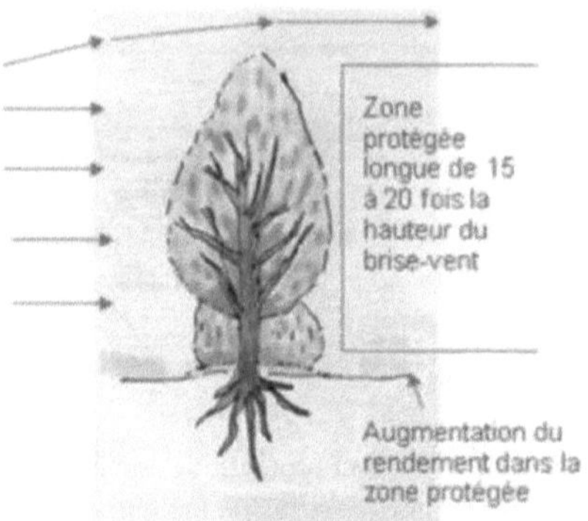

Figure 2: Windbreaks

4.7- Availability of manpower :

In order to guarantee the continuity of work in the nursery, and to obtain quality plant materials that meet the requirements of the market, or the agreed standards for reforestation, the human factor is very important.

In modern or pilot nurseries, despite all are automated, but the value of human labor is required, because all monitoring operations (irrigation, cleaning, transplanting, compost preparation, container stuffing ...) are closely related to man. Manpower must be available, in terms of **quantity**; to carry out all tasks inside or outside the nursery if necessary (collecting bark and branches from trees for composting) and **quality (performance) to** avoid mistakes to be feared in all the different phases of plant breeding.

5- Infrastructure :

5.1- Service Infrastructure :

Forest nurseries are like other companies must be contain all the necessary compartments that the staff working there need in daily life:
- Administration ;
- Reception room ;
- Canteen ;
- Staff Locker Rooms ;
- Sanitary...

5.2- Culture infrastructure :

5.2.1- Culture area :

The development of the cultivation area is anticipated by the earthwork of the ground then by a levelling, the ground was covered with a 20 cm thick layer of material, all-coming type, category 5 to 8 mm; this layer was compacted uniformly over the entire surface with the help of a roller or motor-cylinder while maintaining a longitudinal slope of 0.2% (Case of the pilot nursery of Ouchtata-Tunisia).

Figure 3: Growing area

5.2.2- Composting area :

Is a hardened surface built in cement with specific dimensions, the length is 70 meters, the width is 10 meters, whose surface is 700 m².

The composting area can support up to 15 windrows 10 m long, 2 m wide, 1.5 m high and 1 m spacing between each windrow.

In order not to have to fear any problem of water stagnation in the composting area, it is equipped with a drainage system in the form of two ditches all along the composting area with a slight slope; the role of these ditches is to ensure the evacuation of excess water by a slight slope (Case of the pilot nursery of Ouchtata-Tunisia).

Figure 4: Composting area

5.2.3- Drying area :

The nursery drying area is a hard, cemented surface designed for :
- ✓ **Seed extraction:** the cones of conifers are exposed to sunlight to remove the seeds from the cone either for sale, storage or sowing.
- ✓ **Seed storage:** Requires that the seeds be carefully dried in the shade and in the open air before storage; otherwise the seeds are likely to be a reservoir of germs and moisture, and subsequently become useless.

Figure 5: Drying area

5.2.4- Hangar :
Each nursery should contain a place where all the materials necessary for the production of seedlings in the nursery can be stored:
- Cultivation of plants : Soil, compost, seedling containers, inputs, pesticides
- Equipment : Shade, irrigation equipment...
- Logistics : Tractors, trucks...
- Working equipment: shovels, picks, pruning shears, rakes, scissors, concrete mixer...

Figure 6: Nursery Shed

The hangar is characterized by :

- An opening towards the outside opposite to the prevailing winds, allowing the passage of delivery trucks.
- Be naturally lit during the day and artificially lit at night,
- Be equipped with thermal insulation and a "frost-free" heating system.

5.3-   Research Infrastructure :

The research infrastructure, more exactly the laboratory, must be equipped by

- Baking for drying seeds for sampling ;
- Hermetic chamber for the storage of seeds ;
- Refrigerator used for seed dormancy survey ;
- Disinfected chamber for In vitro culture of seeds before sowing in nursery.

From which the seeds can be analyzed by determining :

- Seed germination rate ;
- Detection of seeds that are attacked by pests ;
- Identification of diseases or insects

<u>Precautions in the laboratory :</u>

In the laboratory, we always handle plant materials, which can be a host to certain troublesome microorganisms (spores, insects ...) unintentionally introduced into the room through objects (sampling materials, packaging ...), or through staff via shoes or clothing, hence the sequelae, contamination of work tools and possibly seeds that test or the operator himself.

Surface disinfection only removes the microorganisms existing in the lab at the time of the operation. Moreover, it does not neutralize all microorganisms and does not prevent recontamination at a later stage.

To achieve optimal disinfection, precautions must be taken to avoid contamination in the laboratory.

- ✓ Disinfect laboratory tools (tongs, scissors...), emerging in alcohol and then expose them to fire to neutralize any contamination effect.
- ✓ Clean and disinfect the benches at least once a day before leaving the lab, also can be done at the end of each handling by **detergents** (Bleach "Sodium Hypochlorite") or **disinfectants** (Ethyl Alcohol "Ethanol 70°", Peracetic Acid, Hydrogen Peroxide "Hydrogen Peroxide" ...).
- ✓ The operator should be dressed in sterile coveralls in the laboratory.
- ✓ Use operational gloves before each handling for one time only and discard them at the end of handling.
- ✓ Sterilize shoes before entering the laboratory.
- ✓ Cleaning and disinfecting agents must be used with caution. The chemical risks during the use of these products can be toxic, irritating, corrosive or flammable and can cause serious after-effects to the operator through contact with the body.

6- Production equipment :

In general, each nursery must have the necessary equipment to be able to produce the plants according to the standards (quantity and quality).

The materials used in nurseries can be divided into two categories:

6.1-   The pots :
6.1.1- Biodegradable pots :

Plants that are grown in biodegradable, root-penetrable pots, also known as "anti-bunting containers", allow aerial ringing of root systems. These containers allow the formation of root apexes ready for planting as soon as they leave the nursery.

Root girdling; **consists of cutting the roots of a tree,** at a certain distance, all around the trunk. Did you know that it is an almost obligatory step when we want to move a tree, in young plants, the ringing is caused by the drying of the root and the inhibition of lengthening of the opening of the root in the open air. This leads to the formation of root shoots at the apexes. The latter resume their development when they are in contact with moist soil during cultivation in the soil,

Aerial ringing is ensured by the use of porous walls into which the roots can penetrate. This type of containers offer the following advantages :

> Increase in the number of secondary roots allowing after planting a better anchoring of the plant and a better supply of water and fertilizer.
> Earlier transplanting is possible, but also improves the preservation of the jars, which allows a more flexible and safer management in the plant.
> Absence of root deformations by winding which allows a better holding of the adult plant and an improvement of its lifespan,
> Very fast recovery after planting

Figure 7: Root appearance in a biodegradable container

6.1.2- Rigid wall pot :

In a pot with a rigid wall that the roots are unable to penetrate the wall of the container. The formation of the root system in a pot depends on the shape of the pot, and the roots rotate against the walls of the pot. This deformation is detrimental because the spiralling of the roots reduces the ability of the roots to develop during transplantation. This has a negative effect on the proper absorption of water and fertilizers, which ends up being strangled; it can limit the recovery and growth potential of plants and the premature death of perennial plants in many cases.

This distortion is called **"Chignon"** which is more serious than the environmental conditions which are difficult. The bun forms during a long period of storage in a container or pot before transplanting in the field. The formation of a root bun is much more common in containers.

Indeed, the plants most concerned by the formation of a bun, those with **strong rooting** cannot be deployed because it is stored in a limited space like a container; plants that **form few branches** and **rotate as they develop**. Among the forest species that are exposed to the bun are: spruce, pine, walnut, giant redwood, fir, oak.

To overcome this disadvantage, there are processes that avoid this root spiraling:

- The biodegradable peat moss bucket for young plants
- The forest bucket for trees with a 20 cm deep taproot allows the taproot not to bend when touching the bottom of a container.
- The above-ground basket for trees and conifers 2 to 3 years old.

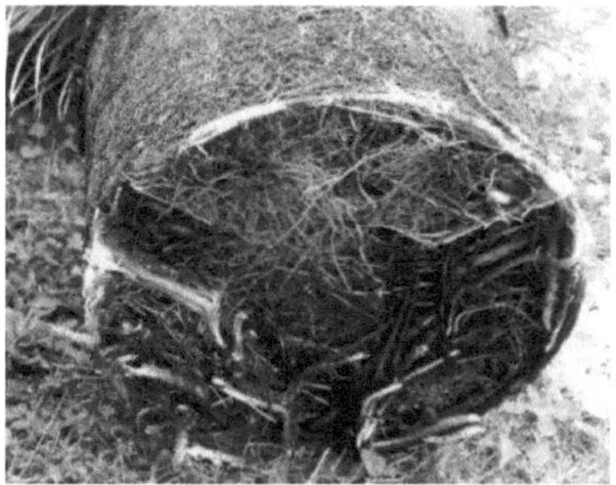

Figure 8: Root deformation in a rigid-walled pot

6.2- Containers :

Containers that are used for growing plants in multi-pot nurseries whose pot sizes are designed in a way that makes them easy to use. The pots, are composed of (15, 50, 40, 130) round or square cells, moulded at the upper end, with holes for the drainage of excess water.

Figure 9: Seedling containers (40 cells)

6.3-   Cemented boards :

This technique is often used in traditional nurseries, it has replaced the cultivation tables in modern nurseries. The boards are in the form of a rectangular cement basin whose base is in direct contact with the soil. The polyethylene bags are filled with potting soil and then placed in the boards. The boards are characterized by a drainage system that allows excess water from watering to be drained off. In addition, they prevent weeds from invading.

Figure 10: Cemented board

6.4-   Boards :

It is a tool made of wood or hard plastic, their handling is manual, it is enough to exert a small force on the potting soil containing the pots to bury the seeds after sowing. The number of seeds deposited in each pot depends on the germination rate.

The manual spreading of silica sand on the containers allows the application of a layer of 3 to 5 millimeters. Excess sand can be removed with a plank and restored in the following containers. The seed should not cover more than 1-2 times the diameter.

Figure 11: Seed burying pad

6.5-   Growing table :

Is an elevated device (cultivation above ground), length is 110 cm, width is 70 cm, the capacity of each board is 10 containers, whose frame is made of galvanized steel. The use of the cultivation tables is during the growing season is to raise the containers to allow **the self-circling of the roots**.

Figure 12: Cultivation table

6.6-   Umbrellas :

The shade is designed to withstand the sun's fiery rays. Shade cloth is a woven net composed of high density flat or round polyethylene yarns treated against degradation caused by the sun's ultraviolet rays. It has a shading degree of 50% which represents the sunny conditions in the undergrowth. Moreover, this percentage of shade is well adapted for the different species that can be produced in Tunisian nurseries. The height is raised 3 meters above the ground.

Figure 13: Shades

6.7- Irrigation system :

The sprinkler system is installed in the shading device. It is provided by sprayers equidistant between them which produces very fine water droplets resembling the fog in the form of circles, thus avoiding the emergence of the seeds or the soil of the containers. The watering system must always be equipped with filters to remove impurities suspended in the water (soil particles, weed seeds.

The irrigation system is linked by a water reservoir, from which water is supplied by rainwater or a borehole. Water circulation is carried out through "PVC" plastic channels and water pumping is done by an electric pump that carries the water from the tank to the sprayers through PVC pipes.

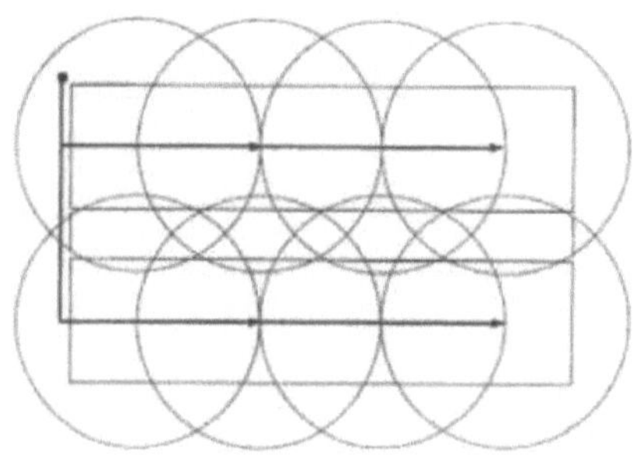

a. Irrigation device - equidistant sprinkler
(Lamhamedi et *al*; 2006)

b. Manual watering - Sprayer

Figure 14: Irrigation system

In some nurseries and especially those that are modern. The fertilization system is integrated into the irrigation system through a soluble fertilizer injector allows the application of fertilizers by sprinklers (fertigation). This technique is called "Fertigation System". This technique is recommended because it allows :

- ✓ Save water use and nutrients ;
- ✓ Maximize the homogeneity of benefiting water and nutrients ;
- ✓ Encourage that all plants receive water and inputs at the same time .

6.8-   Closing :

Nursery fences are used to prevent livestock from entering the nursery can be :

- **Walls:** their construction is expensive, the wall it is effective against the passage of animals, in return it prevents the passage of air and concentrate too intense heat inside the nursery.
- **Sharp hedges:** Trees provide appropriate shade from the sun's rays in the nursery, and also prevent livestock from entering the nursery. For the hedge to be effective, it must connect the trees with barbed wire up to 1.5 to 2 meters and tighten the spacing between them in order to prevent animals of any size from entering the nursery.
- **Grids:** they exist in different heights, with different mesh sizes, protected by galvanization or plasticized, permanently fixed, supported by wooden, metal or cement posts well anchored to the ground.

b. Galvanized Cement Post Grating

a. Galvanized Metal Post Mesh

c. Galvanized live fences

Figure 15: Fencing

6.9- Greenhouses :

To hasten (speed up) the production of plants in nurseries, using the greenhouse culture, usually greenhouses are equipped with metal frames covered by transparent polyethylene designed to save energy. They are mainly found in propagation nurseries.

In addition, greenhouses form a shelter for small plants from frost, protect delicate plants from the cold and the multiplication of plants.

As a precautionary measure, greenhouses should be aired from time to time, removing the polyethylene covering it. Greenhouses must be ventilated from time to time, by removing the polyethylene covering it.

Figure 16: Greenhouses of plants in nursery

References bibliographies :

Anonymous; 1995: Newsletter. The benefits of windbreaks. Ontarion-Canada.

Baltet C.; 1995: The fruit, forest, shrub, vineyard and colonial nursery. Paris, 841.

Christine D. and Isabelle B.; 2014: Surface disinfection in biological laboratories. Institut Nationale de la Recherche et de Sécurité (INRS). France.

Cuissance P; 1984: Multiplication des végétaux et pépinière collection d'enseignement horticole, 177.

FAO; 1992: Forestry in drylands. A guide for field technicians.

Hansen-Catania S. and Miquel M.; 2013: Aerial encircling of root systems in biodegradable root-penetrable pot culture. France.

Lamhamedi M.S., Bertrand F., Luc G. and Christine G.; 2006: Guide pratique de production en hors sol de plants forestiers, pastoraux et ornemental en Tunisie. Direction Générale des Forêts, Tunisia: Pépinière et reboisement. Pampev International.

Minko Mi Obame J.M.; 2009: Preliminary study for the establishment of forest nurseries in Gabon. Polytechnic University of Bobo-Dioulasso-Institute of Rural Development. Burkina-Faso, 14.

Nicolas J. P. and Roche Hamon Y.; 1987: The nursery. Paris, 208.

Ouédraogo A. S.; 1989: Guide technique sur les pépinière forestières. Chad, 116.

Chapter 2 :

Composting

Table of Contents

List of figures

List of tables

General Introduction :

The most important objective of setting up a forest nursery is to produce seedlings in large quantities and of good quality and also capable of withstanding external conditions during reforestation.

In fact, the good practice of the seedlings in the nursery from sowing to transplanting including fertilization, irrigation, maintenance ... etc., not only sufficient to have seedlings with coveted qualities, but there are also other factors no less important that act on the quality of the plants.

To this end, new techniques to improve the quality of plants in forest nurseries, notably **composting** and **mycorrhization, are** proving their effectiveness and are now more widely used in plant breeding.

Party. 1: Composting

1- Introduction :

In nature, waste does not exist. Organic matter is biodegradable and the elements thus released are reintegrated into plants. In our societies, we have long considered organic matter as vulgar waste to be buried, with all the problems that this implies.

To improve the quality of plants in nurseries, foresters often use chemical fertilizers, but they are costly and harmful to nature, causing soil and groundwater pollution if misused. Faced with this situation, foresters integrate another technique based on organic waste from the forest (branches, leaves, bark...), in order to improve the productivity of seedlings and serve their food on the one hand and soil fertility in forest nurseries on the other, this technique is called **"composting"**.

Composting is a green technology that replaces chemical fertilizers used in nurseries and allows the valorization of residual organic matter. Plants planted in an environment containing compost are stronger in terms of resistance to climatic hazards and have a better yield than other plants that grow in an ordinary environment.

2- Composting :

2.1-   Definition :

Composting is defined as the decomposition of biodegradable materials. The stabilization of organic substrates in conditions that allow the development of thermophilic temperatures, the result of a heat production of biological origin, with the obtaining of a final product sufficiently stable for storage and use on the soil without negative impacts on the environment.

Composting is a process in which organic materials are biologically degraded under controlled conditions to finally produce a humic-rich substrate called "compost". Organic compost is obtained through a process of mineralization and humification.

Composting is the process of aerobic transformation of fermentable materials under controlled conditions. This process results in a stable, fertilizing material, rich in humic compounds, which is used as an organic amendment to improve soil structure and fertility.

Composting is a controlled process of degradation of organic constituents of plant and animal origin, by a succession of microbial communities evolving in aerobic conditions, leading to a rise in temperature, and leading to the development of humified and stabilized organic matter. The product thus obtained is called compost.

Compost is the result of the composting process. Compost is a dark, blackish-brown substance. It is humus containing living organisms and minerals that can be used to improve the structure and fertility of the soil and to provide food for plants.

Composting is a process by which biodegradable materials are put together to be converted into a stabilized humic amendment, through the work of living biological organisms under controlled conditions".

- Composting is a natural process!
- Compost is a natural product!

In summary: Composting is a process of degradation of organic matter biologically through the action of living beings (bacteria, fungi, earthworms ...) under controlled conditions, to transform organic waste into a sufficiently stable substrate rich in humus, used to fertilize the soil, without negative impact, this is the "compost", to which may have been added certain enriching products such as: phosphate. The genesis of compost is obtained by a process of mineralization and humification which is accompanied by a calorific production of biological origin. This process takes place in the presence of moisture and oxygen.

2.2- Advantages of composting :

- ➢ Compost can be used as a **biological control method to** reduce the use of pesticides in the nursery.
- ➢ Composting is accompanied by the release of heat and gas, mainly carbon dioxide if aeration is sufficient, which allows the **destruction of pathogens.**
- ➢ Composting can **reduce masses and volumes by up to 50% compared to the initial waste,** due to the mineralization of organic compounds, water loss and the modification of the porosity of the environment.
- ➢ The rise in temperature and the release of inhibiting agents allows the **destruction of weed seeds.**
- ➢ The porosity and fibrous texture of the organic compost favours a **good development of the root system;**
- ➢ **High water retention capacity** used for the plant;
- ➢ **High cation exchange capacity** to retain and make available the nutrients necessary for plant growth;

2.3- Limits and constraints of composting :

- ➢ Availability of plant material, animal dung and water;
- ➢ Transportation problem in case the composter is far from the place of use ;
- ➢ Labour-intensive work (especially digging pits) ;
- ➢ The pits require regular annual maintenance;

2.4- Composting Characteristics :
2.4.1- Genesis of heat :

In general, if the quantity of material is sufficiently degradable and if the ventilation and humidity conditions are favourable to the development of micro-organisms, the biological activity of the first days generates an increase in temperature, from which values can reach 60 to 70 ° C. Then the temperature stabilizes for a few days to a few weeks depending on the operation performed, then it decreases regularly until it reaches a fixed value. This drop in temperature, especially rapidly it would be a forced ventilation, reflects the slowing down of biological activity.

2.4.2- Smell :

Odor emissions may occur during storage of raw waste prior to treatment. It is caused by the presence of organic molecules such as fatty acids, aldehydes, ketones, nitrogen compounds (such as ammonia and pyridine) and sulphur compounds (such as mercaptans and sulphides). These unpleasant odours diminish during the composting process and tend to disappear at the end of treatment. They are then replaced by the "smell of humus".

Unpleasant odour emissions may persist during treatment, due to either low ammonia emissions or poor ventilation leading to the development of anaerobic reactions.

2.4.3- Loss of mass and volume :

A loss of mass and volume, due to the loss of CO_2 and H_2O by the evaporation of water under the effect of heat.

2.4.4- A homogeneous structure :

In a compost one cannot distinguish the different plant debris under the effect of biological degradation, consequently one obtains a compost of the same color, the same texture.

2.5- Composting technique :
2.5.1- Composting on the ground :

This technique is the simplest. It consists of leaving crop residues on the soil (leaves, stems and roots) and in the soil (for the roots) where they are then decomposed by insects and soil microorganisms. This technique is often used in agriculture, especially in cereals.

2.5.2- Compost in heap :

This is the ideal technique in nurseries. This method consists in arranging its waste in heaps (0.5 to 1.5 m high) in a shady place sheltered from the wind. Be sure to mix the waste well with the substrate at each feeding. When the pile (stack) reaches a large volume, turn it over (stir with a fork) to re-mix the additions. This is also an opportunity to control moisture.

2.6- Type of composting :
2.6.1- Aerobic composting :

This type of composting is characterized by the **presence of oxygen** in (abundance/profusion). During this process, **aerobic micro-organisms** decompose the organic matter and produce carbon dioxide (CO_2), ammonia (**NH4**), water (H_2O), **heat** and **humus**, which is the relatively stable final organic matter product.

Although **aerobic composting** produces **certain organic acids** (intermediate organic compounds), these are then broken down by aerobic microorganisms. Thus, the resulting compost is **relatively unstable** organic matter and carries **very little risk of phytotoxicity**.

The heat induced during the composting process acts as a catalyst to decompose proteins, fats and complex sugars such as cellulose and hemicellulose, it can thus reduce the composting time, and it also destroys microorganisms, which are pathogens for humans and plants.

This type of composting is most commonly used in the formation of compost for use as an organic amendment to improve soil structure and fertility and as a food source for nursery plants instead of chemical fertilizers.

2.6.2- Anaerobic composting :

On the contrary, this process takes place when **oxygen is lacking** or absent. **Anaerobic microorganisms** dominate and produce intermediate compounds such as methane (**CH4**), organic acids (**R-COOH** carboxylic acids), hydrogen sulfide (H_2S) and other substances.

In the absence of oxygen, these compounds accumulate and are not metabolized. Many of these compounds have strong odors and some of them have phytotoxicity.

In addition, anaerobic fertilization is a process that occurs in conditions where the temperature is low and the seeds of weeds and pathogens are not affected. In addition, it often requires more time than aerobic composting. All these disadvantages outweigh (offset) the advantages of this process.

Although nitrogen and potassium nutrients are lost in greater amounts in aerobic and anaerobic composting. Aerobic composting remains more efficient and useful than anaerobic composting for improving plant quality in forest nurseries.

2.7. Stages of composting :
2.7.1- collection of plant material :

In this step, all materials destined for composting are collected such as (leaves, branches, bark, roots...etc.). It does not need skilled labor hands just workers and materials needed for the collection.

- Disconnection: pruning shears, hatchet ;
- Collection of green materials lying on the ground: pallet ;
- Collects the raw material: bag.

2.7.2- Sorting :

After harvesting the raw materials, workers must manually sort the harvested product to separate the non-degradable materials can be collected with the plant materials (plastic, gravel ...) to the degradable materials. This technique is carried out on a cemented surface (e.g. drying area) so that they are suitable for crushing and do not interfere with the blades of the crusher, and finally have a purely organic substance.

2.7.3- Grinding :

The composting raw material must be shredded by a shredder equipped with a "knife-to-hammer" system that facilitates the degradation of the organic matter and then obtain particles/chips at optimal dimensions. Once the branches have been cut, they must be quickly crushed to prevent any loss of moisture in the foliage.

It is required to use a shredder with 20 horsepower, also each nursery must have a minimum of two shredders to improve the shredding efficiency of the green material.

It has been proven that the addition of ammonium nitrate (NH_4NO_3) to the chipper's "knife-hammer" system considerably reduces the composting period for bark and forest biomass. In fact, mature compost can be obtained after a period of 105 days of the composting operation.

2.7.4- Windrowing :
2.7.4.1- Swath size :

After the (shredding/grinding), the organic material is placed in the form of windrows. The dimensions of each windrow are as follows, height (h) is 1.5 meter, width (W) is 1.5 meter and length (L) is closely related to the composting surface (composting slab).

2.7.4.2- Windrow texture :

The texture of the windrows can be homogeneous (same component, either branches, bark or leaves) or heterogeneous (branches and leaves). When windrowing, it is preferable to use piles with a heterogeneous rather than homogeneous texture because the woody material comes from branches and the non-woody material comes from leaves, bark and roots (cellulose and hemicellulose) which is rich in mineral elements. This makes composting more fertile in terms of organic and mineral composition since it replaces chemical fertilizers.

2.7.4.3- Windrow preparation :

The formation of swaths is done by successive layers of crushed particles based on leaves, branches, bark or roots..., each layer is 20 centimeters thick. A layer 20 cm thick represents a volume of 0.3 cubic meters per linear meter of swath. This technique facilitates the construction of swaths, the addition of ammonium nitrates and their moistening.

The addition of ammonium nitrate (NH4NO3) is used to moisten the andin and is done as the quantity of material is crushed. For each cubic meter of shredded material (e.g. acacia or bark), an average of 3.0 Kg of ammonium nitrate must be added in two applications.

The first and second applications are carried out during windrow construction and the first turning with a watering can, respectively. The ammonium nitrate was diluted in an aqueous solution at a rate of 1.5 Kg/ 20 liters of water/m3.

The quantity of water (20 liters) can be adjusted according to the humidity of the grinded material and the ease of application. The swath humidity should reach 50 to 70%. In addition to the addition of nitrogen, the moisture content of the mulch (50-60%) can be increased.

Nitrogen and moisture are two important factors for the multiplication of microorganisms and consequently the degradation of polymers (cellulose, hemicellulose and lignin) and the reduction of composting time.

2.7.5- Turning the windrows :
2.7.5.1- Temperature effect :

The success of composting is based on the daily monitoring of the temperature and the evolution of this factor according to the different phases of composting and is a reliable indicator of the decomposition of organic matter. The daily temperature will be the average of the temperatures measured always at the same time of day, at four or six representative locations in the windrow.

In order for composted organic matter to degrade, the temperature of the heap must be between mesophilic and thermophilic, i.e. the thermal value of the windrow varies between 20 and 70 °C. The turning of the windrow will take place when the temperature depreciates by 5 to 10°C so that the temperature rises to 70°C.

Indeed, several turnarounds must be performed. Generally speaking, 5 to 7 turnings are applied over a period of 90 to 120 days before the compost has reached maturity.

After each turn, the temperature gradually decreases until it is equal to the temperature of the ambient atmosphere. This is where the compost has become mature.

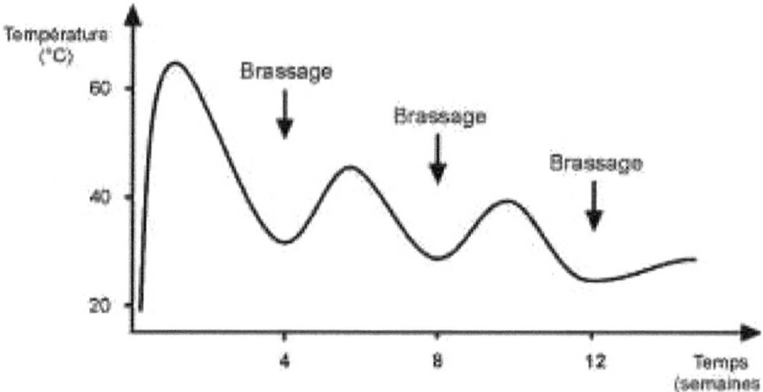

Figure 17: Effect of regular turning on compost temperature

2.7.5.2- Moisture effect :

The monitoring of moisture content in the swath is no less important than the temperature. For this purpose, regular monitoring is required to maintain an optimal percentage of water for a good compost.

This test consists of taking samples from different representative parts of the swath, using the hand to exert pressure on each sample if they have shown signs of dryness (dry sample due to low humidity) due to the absence of water released during the pressure. To overcome this problem, the swath must be moistened to the point where the humidity must be maintained in the range of 50 to 70°C.

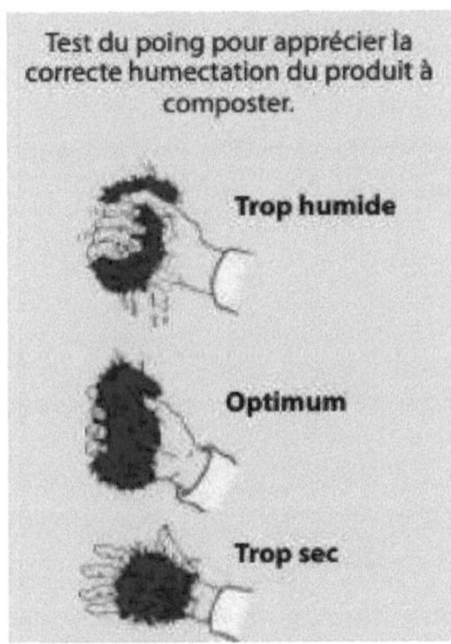

Figure 18: Hand moisture test

2.7.6- Sifting and storage :

When the composting process has been completed, and the final product is now mature and ready for use in the nursery, the nurseryman is required to screen the compost to separate the large particles from the small ones either for use or for storage and use later.

2.7.6.1- Sieving :
To get the most out of your compost, sift it! This will allow you to get the highest quality compost from your composter.

Before using the compost it is recommended to screen the compost to separate the fine from the coarse elements, as the fine elements decompose faster than the coarse ones. Finer compost is more easily used for sowing or repotting, by mixing/blending it with soil and potting soil to remove branches and undecomposed material.

The sieve separates coarse to fine elements. The sieve residues (those that fall on the front) can be recycled later.

Figure 19: Compost sieve

2.7.6.2- Storage :

This is the last phase of the compost life cycle. The excess quantity must be kept in a place with conditions that favour the maintenance of its quality.

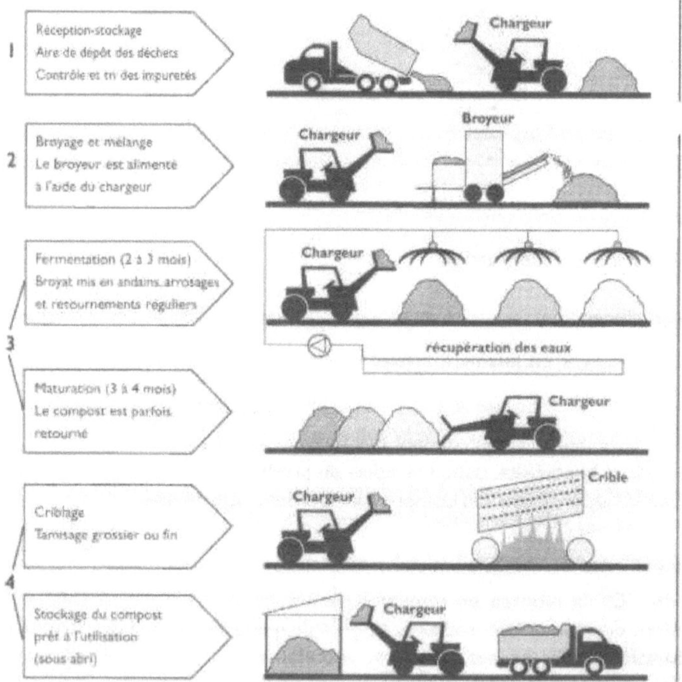

Figure 20: Composting stage (ADEME)

2.8-   Composting process :
2.8.1- Mesophilic phase :

This is the initial stage of composting. The micro-organisms are replaced by others that prefer lower temperatures, especially the bacteria and fungi that develop during this phase and invade the raw materials. It is no longer necessary to turn the compost, but it must remain aerobic.

During this phase, oxygen is very important since bacteria multiply strongly and microorganisms break down simple molecules such as simple sugars, lipids, amino acids and transforms part of complex molecules (or polymers) such as (proteins or starch, pectins, hemicellulose, cellulose ...). This activity under the action of microorganisms causes an increase in temperature in the windrow from 10-15°C up to 30 to 40°C in a few days even after turning over the pile, moreover a release of water vapor and CO2 hence the decrease in the carbon/nitrogen ratio. Cellulose degradation during this phase is responsible for more than 75% of the dry weight loss.

Unmatured compost may have higher concentrations of organic acids, high C/N ratios, extreme pH values or high salt content. All of these characteristics can damage or destroy plants.

2.8.2- Thermophilic phase (Active) :
In this phase the decomposition of materials occurs most rapidly. Once the different elements are mixed and stacked, the temperature reaches a maximum (60 to 70°C) in the windrow. Decomposition is then carried out by thermophilic and aerobic organisms, including bacteria, actinomycetes, fungi and protozoa. The microorganisms use oxygen to consume raw materials and produce carbon dioxide. Raising causes the compost to be disinfected and diseases or seeds present in the compost are annihilated, so we talk about compost sanitation.

At the end of the phase, the compost will have lost nitrogen in the form of ammonium ($NH4+$), which can be volatilized as ammonia ($NH3+$). Water evaporation is more important during this phase. The release of CO_2 can lead, at the end of the thermophilic phases, to a loss of up to 50% of dry weight.

This phase ends when the temperature no longer increases, even if the oxygen supply is increased. This means that the simple compounds are depleted, hence the stability point is reached, i.e. the peak or pasteurization temperature is reached in the windrow. The compost is therefore said to be "raw".

The active phase can last from 1 to 4 months if the windrow is turned frequently, from 4 to 8 months if it is turned less frequently, and from 6 to 24 months in the case of passive composting without turning or active aeration.

2.8.3- Cooling phase :

It is the intermediate phase between the thermophilic phase and the maturation phase. This phase begins when the bacteriological activity decreases due to a lack of easily degradable organic matter.

The compost then slowly loses heat and goes down to room temperature; this drop in temperature favors the colonization of the environment by new mesophilic actors such as fungi, the macrofauna composed of earthworms, sowbugs, manure worms and actinomycetes which will degrade the polymers (cellulose, hemicellulose and lignin).

During this phase, the compost loses its bad smell, and humus begins to be formed through the incorporation of nitrogen into the complex molecules of the compost. This phase will be quite slow and can last up to a month.

2.8.4- Maturation phase :

This is the last phase in the composting process. This stage is characterized by a little microbiological activity in particular the activities of macrofauna (earthworms) which deals with this phase and not microorganisms. The organic materials are **stabilized and humified** compared to the raw materials put to compost, because the organic matter that made up the compost is simplified into mineral substances.

The temperature is about 30°C during this phase but it continues to decrease to that of the outside environment, and the compost has become stable, which means that it can no longer decompose. This phase of composting is the slowest but its duration varies depending on the types of materials that were put into the compost from the beginning.

The degree of maturity is an indication of the degree of humification, or conversion of organic matter into humic substances resistant to microbial decomposition. Several tests can be used to measure the degree of maturity of the compost. Laboratory analysis or germination tests using lettuce or watercress seeds can be used. If the compost is not ready, it will damage the germinating seeds and kill the plants.

The degree of maturity is an indication of the degree of humification or conversion of organic matter into humic substances resistant to microbial degradation. Several tests are used to measure compost maturity. Laboratory tests or germination tests can be used with lettuce or watercress seeds. If the compost is not ready, it will damage the germinated seeds and kill the plants.

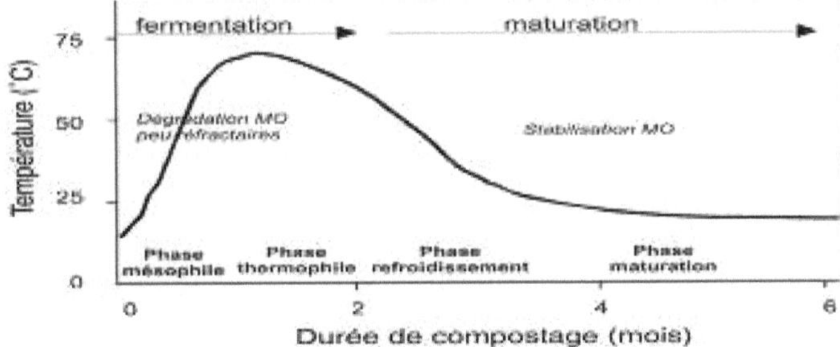

Figure 21: Composting process

2.9- Method for assessing compost maturity :

2.9.1- Empirical (experimental) method :

This involves observing and manipulating the compost to determine if it is mature. A mature compost has the following characteristics:

- ✓ It has no ammonia smell;
- ✓ Its temperature is low even if the humidity remains good;
- ✓ It is granular, dark and smells good;
- ✓ There are at least three types of arthropods: sowbugs, ground beetles and centipedes;
- ✓ The original compounds can no longer be distinguished with the naked eye.

2.9.2- Physico-chemical method of composting :

2.9.2.1- Physical method :

2.9.2.1.1- Aeration test :

Aerobic composting requires large amounts of oxygen, especially in the initial phase. Ventilation is a source of oxygen and is therefore an indispensable factor in the decomposition of organic matter. When the oxygen supply is lacking, the growth of aerobic microorganisms remains limited, which subsequently slows down decomposition and triggers the anaerobic process that produces unpleasant odours.

In addition, ventilation reduces excess heat and eliminates water vapor and other gases trapped in the heap/tap. Heat removal is particularly important in hot climates due to the high risk of high temperatures and fire. Therefore, good ventilation is necessary for effective composting.

To keep a good oxygenation, turnings are important. They will allow the materials to be mixed so that they are all well attacked and to maintain aeration, which reduces the causes of overcrowding.

2.9.2.1.2- Moisture test :

Humidity is necessary to ensure oxygen exchange and thus the activity of living beings. The ideal humidity is the one that allows a good compromise between the two important parameters for microbial activity, i.e. the aeration of the environment and its water content.

13

In practice, the moisture content of the material to be composted should be between 50 and 80%. This percentage depends on the nature of the waste being composted, so that the moisture content at the end of the process reaches 30%. If the windrow is very dry, it slows down the microbial development, so the process is slower, while the humidity is exceeded, saturates the lacunar spaces and suffocates the micro-organisms, the environment has become anaerobic.

The moisture content of the compost matrix conditions oxygen exchange and therefore microbial activity. The ideal moisture content is that which allows a good compromise between the two parameters important for microbial activity, namely the aeration of the medium and its water content.

2.9.2.1.3- Temperature test :

High temperatures characterize **aerobic composting** processes and are important indicators of microbial activity.

The composting process is divided into four main temperature ranges, heating (mesophilic and thermophilic), pasteurization, cooling and maturation. Turning and ventilation of the windrows can be used to regulate the temperature.

i. Mesophilic phase: The ideal temperature for the initial composting stage is 20 to 45°C, due to the decomposition of soluble elements,
ii. Thermophilic phase: heat-loving organisms may have controlled the following steps, but the temperature between 60 and 80°C is ideal.
iii. Pasteurization phase: this is the peak temperature, the organisms (fungi) that are responsible for the degradation of soluble compounds (sugars, starch and lipids) are destroyed. During this phase the temperature varies between 55 and 65°C. Pathogens are generally destroyed at 55°C and above, while the critical point for weed seed removal is 62°C.
iv. Cooling phase: "Actinomycetes" type fungi settle and promote the degradation of polymers (cellulose and hemicellulose). Once the temperature stabilizes by approaching that of the ambient atmosphere, it is the maturation phase characterized by the formation of humus.
v. Ripening phase characterized by the formation of humus and that the temperature in the swath is almost equal to that of the outside environment.

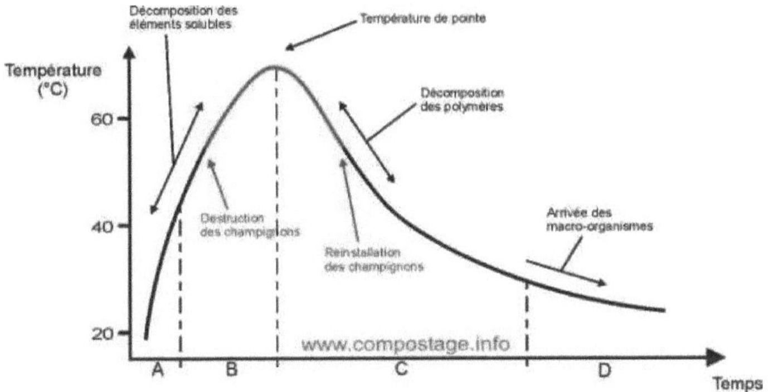

Figure 22: Temperature variation during the composting process

2.9.2.1.4- Sieving test :

When the composting process has come to an end, the final product is subjected to a sieving or screening test, which allows it to identify the maturation of the compost.

The compost is placed in a sieve with a screen size of 25 mm. If most of the sieved compost passes through the sieve, the compost is ripe. The method often indicates over-ripe compost with little fertilizer value. The method is not valid for composts that have been previously shredded.

2.9.2.1.5- Porosity test :
Principle :

Porosity represents the spaces that are not occupied by particles. In general, these spaces are filled by water and/or air; it is a storage place for the irrigation water needed to feed the plants and the air for gas exchange at the roots.

Soil porosity is defined as the ratio between the total volume occupied by pores (voids) and the total volume (voids + solids) of a sample.

The porosity of a substrate depends on the characteristics of the soil matrix (texture, bulk density), biological activity (presence of galleries due either to earthworms or roots) and cultural practices, the characterization of porosity is very useful in the diagnosis of a compost.

Aerobic micro-organisms consume oxygen to oxidize the organic compounds they feed on. The compost heap must contain at least 30 to 35% of gaps to allow good air circulation.

The **"Beer kan"** infiltration test consists of measuring the rate of water infiltration into the soil, in the case of wet and wiped/wet soil. A determined volume of water is poured into a cylinder sunk into the surface of the soil. The time required for complete infiltration is noted for one volume of water poured. The operation is repeated until the infiltration time stabilizes.

Porosity (Φ) is a numerical value defined as the ratio between the volume of voids and the total volume of a porous medium: Through this test, it is possible to quantify the type of space and the volume available to manage these two elements. Porosity is expressed as :

$$\Phi = V_{pores} / V_{total}$$

Three types of porosity can be identified :

- **Macroporosity:** relative to pores with a diameter greater than 50 nanometers.
- **Meso-porosity:** relative to pores with a diameter between 2 and 50 nanometers;
- **The microporosity:** relative to pores whose diameter does not exceed 2 nanometers.

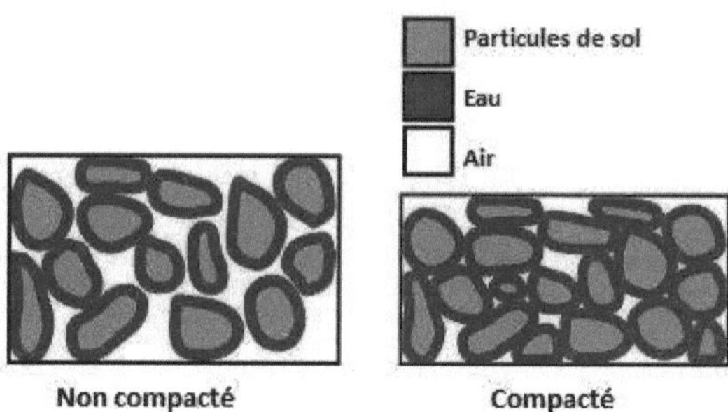

Figure 23: Substrate porosity

Objective :

The objective of this test is to evaluate the rate of water infiltration of the compost samples tested in order to take appropriate measures to obtain a growth substrate with optimal porosity that promotes good plant growth.

This test indirectly provides information on soil porosity, which can be cross-referenced with other indicators of soil fertility (physical, chemical, biological).

Measuring equipment :

The method is easy to implement because the necessary equipment is common and inexpensive. In addition, the duration of this test is limited in comparison to conventional methods. It is directly related to the porosity of the soil (i.e. the infiltration rate) and the number of times the test is repeated.

For this test, it is required to have

- A 1.5 liter plastic bottle with cap
- Cylindrical mug in PVC with a diameter of 30 cm and a height of 15 cm to be measured.
- The bottle should be cut in half and then placed one inside the other, so that the part with the cork down is in the other half of the bottle.
- Afterwards, 500 ml of water must be measured and put into the funnel-shaped part of the bottle (with the cap) and draw a line indicating the exact measurement of the 500 ml. This line will be the reference point for the different measurements of the compost and/or substrate samples.

- The compost or substrate to be tested must be completely dry. If the material is not completely dry, the values obtained will be erroneous. The following steps must be carried out in order.

Methodology :

i. Fill the part of the funnel bottle with the material to be tested up to the previously marked line (reference line) ;
ii. Moderately shake the bottle to obtain material compaction similar to that of a substrate in a culture container. If the material does not arrive directly on the line, add or remove what is necessary (check that there is material in the bottle cap) and level the surface with your fingers. Proper placement of the material is important to obtain accurate values.
iii. Fill the measuring cup with 500 ml of water (Total volume of the container: VT).
iv. Pour this water into the section of the bottle containing the material until water appears on the surface of the material.
v. Keep aside the water that will remain in the cup as it will be needed later.
vi. Leave the water in the bottle with the material for approximately one (1) hour to allow the water to properly penetrate the compost or substrate.
vii. After one (1) hour, check that the water is still on the surface, otherwise add water until it appears on the surface. For this operation, use the water remaining in the measuring cup from the previous step.

Measurement Parameters :

To evaluate the porosity of the material, you must use the three parameters measured previously :

- **VT : Total volume of the container ;**
- **VA: Volume of water required to saturate the material ;**
- **RV: Volume of water collected by the drainage of the equipment.**

i. Calculate the total amount of water used to saturate the material. For example, if there is 275 ml of water left in the measuring cup, the amount of water used is: 500 ml - 275 ml = 225 ml (Volume of water used to saturate the substrate: VA). At this point, the measuring cup can be emptied.
ii. Open the bottom cap of the compost bottle and allow water to drain into the bottom container. Let drain for 10 minutes.
iii. Measure the amount of water that has drained from the substrate using the measuring cup: e.g. 150 ml (Residual water volume: VR).

Example :

In our example, we have the following values:
- **VT = 500 ml**
- **VA = 225 ml**
- **RV = 150 ml**

With these values, the three types of porosity can be calculated:
- **Total porosity (PT) = (VA ÷ VT) X 100**
- **Macroporosity (Ma) = (VR ÷ VT) X 100** : Air spaces
- **Microporosity (Mi) = PT- Ma:** Spaces occupied by water -

With the values of our example, the calculated porosities would be :
- **PT = (VA÷ VT) X 100 = (225 ml / 500 ml) X 100 = 45%.**

- **Ma = (VR ÷ VT) X 100 = (150 ml / 500 ml) X 100 = 30 %.**
- **Mi = PT - Ma = 45% - 30% = 15% Mi = PT - Ma = 45% - 30% = 15% Mi = PT - Ma = 45% - 30% = 15**

The recommended porosity values for an optimal growth substrate are approximately :
- **Total porosity (PT) 50 - 60%.**
- **Macroporosity (Ma) 25 - 30%.**
- **Microporosity (Mi) 20 - 25**

2.9.2.2- Chemical method :

2.9.2.2.1-   Acidity test (pH) :

The pH has a considerable effect on the multiplication and life of microorganisms. To have a compost with good physico-chemical characteristics, rich in organic matter and nitrogen and excellent water retention. Moreover pH must not exceed 8. The pH has shown some variation according to the four phases of composting (mesophilic, thermophilic, cooling and maturation).

i.   **Acidogenesis phase:** the pH undergoes a decrease due to the production of carbon dioxide and organic acids generated by the microorganisms (Meso and thermophilic phase; A).

ii. **Alkalization phase:** the pH increases and exceeds 7 and the medium becomes more basic due to the production of ammonia (pasteurization phase; B).

iii. **Stabilization phase:** the pH stabilizes and is accompanied by a decrease in the C/N ratio and slower reactions (cooling phase; C) ;

iv. **Neutrality phase:** The pH becomes more stable and approaches neutrality between 6 and 8 5 Maturation phase; D).

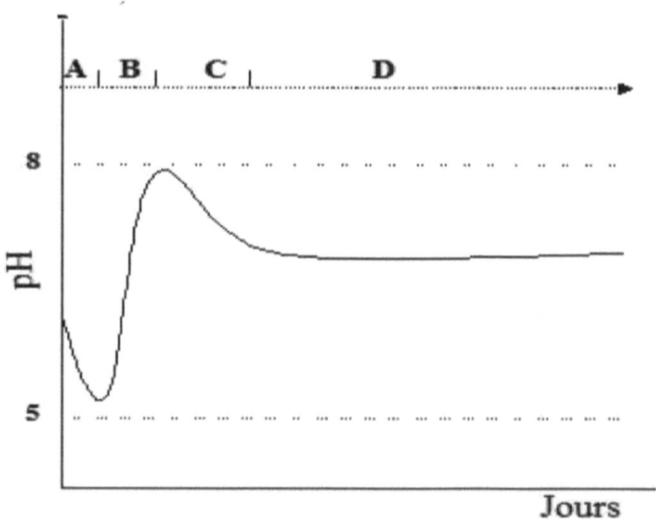

Figure 24: Curve of pH change during composting (adapted from Mustin 1987 and Dalzell et al. 1988).

18

2.9.2.2.2- Carbon content :

Nitrogen is, along with carbon, one of the most important elements in the constitution of organic matter.

Organic carbon is a major component of composted organic waste. Total carbon is composed of total organic carbon (TOC) and mineral carbon in the form of carbonate and bicarbonate. TOC generally accounts for more than 90% of the total carbon in compost.

TOC content decreases during composting. The main reason for this decrease is the use by micro-organisms for their metabolism, resulting in carbon dioxide (CO_2) mineralization.

2.9.2.2.3- Nitrogen content :

The multiplication and growth of microorganisms is proportional to the amount of nitrogen available in the compost. The biomass used should be relatively rich in nitrogen (leaf biomass and other green residues), a key element in the growth of microorganisms that degrade cellulosic compounds.

During the composting process, the organic nitrogen in the composted waste becomes mineralized. The main forms of mineral nitrogen in composts are ammonium (NII4+), and nitrates (NO3-) when the nitrification process is complete.

Part of this mineral nitrogen is reintegrated into the microbial metabolism of the active microorganisms during composting. Part of this mineral nitrogen is integrated into the organic matter of the composts during their humification, and part is released into the matrix in the form of mineral nitrogen.

At the end of fertilization, the mineralization phenomenon has become predominant, and an increase in NO3- content is often observed. Consequently, an increase in the total nitrogen concentration was observed in the residual dry matter. However, nitrogen losses are possible either through nitrate leaching in the case of compost windrows that are not protected from the weather, or through the volatilization of ammonia (NH3) or nitrous oxide (N2O).

Total nitrogen generally represents 1 to 4% of the total dry mass of compost, and is composed of less than 10% mineral nitrogen.

Table 1: Source of Compost Components

Carbonaceous waste	**Nitrogenous waste**
- Shredded branches, dead leaves, straw (these materials should be stored so that they are always available for mixing with nitrogenous materials); - Biodegradable litter from herbivorous animals	- Vegetable waste, (hedge trimming, lawn mowing...), green leaves,

2.9.2.2.4- C/N ratio :

Microorganisms need carbon (C), nitrogen (N), phosphorus (P) and potassium (K) as their main nutrients. To make compost, it is important to pay attention to the amount of carbon and nitrogen supplied.

The C/N ratio is a particularly important factor in composting. Indeed, the micro-organisms responsible for fermentation need 15 to 30 times more carbon than nitrogen to grow and decompose organic matter.

It is widely recognized that the biodegradability of organic waste is dependent on the C/N ratio. Several studies have identified optimal values and limit values for composting.

According to **Larsen and McCartney (2000),** they found that for optimal composting, the appropriate C/N ratio is between 25 and 30, although it is the ratio between 20 and 40 that is also suitable for satisfactory composting.

If the C/N ratio is higher than 40, the growth of microorganisms is limited and means a longer composting time. A C/N ratio below 20 results in an under-utilization of nitrogen, then excess nitrogen can be lost to the atmosphere as ammonia or nitrous oxide, which can cause an odour problem.

C/N decreases during the composting process to values generally between 8 and 25 at maturity.

2.9.2.2.5- Polyphenol :

For compost to mature and be used as a beneficial substance for plant growth and soil structure improvement, **it must not contain phenol-based products.** Some phenols (aromatic alcohols produced by plants) have important biological functions (biochemical defense against microbes and fungi in plants in particular) in certain species, but they are toxic or highly toxic to humans and other species, and can also affect the germination of forest seeds. When abnormally disseminated in the environment, phenols are air, soil or water pollutants.

Polyphenols include hydrolyzable and condensed tannins. **Soluble** condensed and hydrolyzable tannins react with proteins and reduce their microbial degradation. **Insoluble condensed tannins** bind cell walls and proteins and make them physically and chemically less accessible to decomposers.

Polyphenols and lignin attract more attention as inhibiting factors. To this end, the contents of these two substances are used to classify organic materials in order to achieve a better use of natural resources on the farm, including composting.

2.9.2.2.6- Lignin content :

Lignin is one of the main components of plant cell walls and its complex chemical structure makes it highly resistant to microbial degradation. The nature of lignin affects two to two facades.

i. Lignin reduces the bioavailability of other cell wall components, resulting in a lower C/N ratio than commonly reported.

ii. Lignin acts as a porosity enhancer, creating favourable conditions for aerobic composting. Therefore, while the addition of lignin decomposing fungi can in some cases increase the available carbon, speed up composting and reduce nitrogen losses, in other cases it can lead to a risk of increasing the actual C/N ratio and poor porosity, two factors behind the extension of composting time.

2.9.3- Biological method :
2.9.3.1- Germination test (Who; 1987) :

A germination test is carried out using a large quantity of representative samples of compost to ensure its maturity. This test must be performed before using the growth substrate.

Methodology :

- Fill 15 jars with different soil and compost mixtures (example: volume ratio of 0:1, 1:3, 1:1, 3:1, 1:0).
- Moisten the substrate before sowing the seeds: Watering should be carefully monitored to maintain a certain moisture content of the compost around each seed to promote germination.
- Plant 30 grains of barley per pot and ideally do 3 repetitions.
- Cover the mixtures with 200 ml of sand (1-1.5 mm) and allow to germinate under identical conditions.
- Harvest after 2 weeks and count the number of germs.

Interpretation of results :

Results with soil without compost are considered 100% (treatment without compost = control).

- **germination % = 100 * $N_{T/30}$ * $N_{C/30}$**
- N_T : Number of seeds germinated in the treatment
- N_C: Number of seeds germinated in the control

Subsequently, it will be possible to evaluate the germination rate, and the initial seedling increases for each of the different composts. Thus, good seed germination means that the compost is ripe, non-toxic to seeds, and ready to be used for the production of forest seedlings.

2.9.3.2- Test plan : (Spohn; 1969) :
Methodology :

- Fill a 38 x 28 x 6 cm tray with compost.
- Spread 10 g of watercress seeds on the surface.
- Cover with a glass plate for the time of germination
- Weigh the tray every two days and replace the water that has evaporated.
- Cut and weigh the watercress after 5 days.

Interpretation of results :

- If you get 60 to 100 g of watercress, the compost is of good quality.
- If the compost is of poor quality, the seeds will not germinate or will give few shoots.
- If the compost is unripe or putrid/poor quality, the cress will have an unpleasant aftertaste.

Comments :

- The fastest results are obtained with garden cress but other plants can be used. Barley, for example, is very sensitive to phytotoxicity.
- The test is approximate but indicates the most important characteristic of a compost: the absence of phytotoxicity.

3- Recommendations :

3.1- Biomass related :

3.1.1- Wood products (branches) :

- Branches used as base material are taken from a young stand with abundant leaf biomass.
- Branches should be 2 to 3 cm in diameter, if the supply of branches is abundant, and can be up to 7 cm if the quantity of branches is insufficient to increase the life of the knives and hammers in addition to reducing the carbon content in the initial shredded material.
- The size of the shredded particles is a very important factor in composting :
    i. The size of the materials composted should be small, from 1 to 3 cm (the surface area attacked by microorganisms is much larger), and consequently the composting time will be reduced.
    ii. The too fine particle is not recommended (affects the aeration by reducing the lacunar space: porosity), moreover slow decomposition of the organic matter, is accompanied by a release of disgusting/noxious smells (rot / rot).
    iii. The coarse particles, reduces the surface of attack by microorganisms, thus the composting time increases.

3.1.2- Green products (leaves) :

- The biomass of the green matter or the volume of the foliage should make up 25-50% of the mixture.
- the woody material should vary between 50 and 75%, this proportion must always be adapted according to the materials used.
- The foliage must have a dark green, young color.

3.2- Related to chemical properties :

- The use of branches with old, tough and slightly chlorotic foliage partially disadvantages the degradation process because of its low nitrogen content.
- The biomass used should be relatively rich in nitrogen (foliar biomass and other green residues), which is essential for the growth of microorganisms that degrade cellulosic compounds.
- The carbon-nitrogen ratio should be between 25 and 35.

3.3- Linked to the construction of the windrow :

- The size of the swath has a significant effect on temperature variation and oxygenation. In order to reduce the dissipation of heat loss from the windrow and maximize oxygenation, windrows of 1.5 m wide by 1.5 m high and length are recommended depending on the composting slab.

3.4- Related to the composting process :

-   If the material has a high C/N ratio (little foliage or old foliage), nitrogen enrichment is required at the time of windrowing.
-   The daily temperature will be the average of the temperatures measured at the same time of day at four or six representative locations in the windrow using a 90 cm long thermometer.
-   To reduce the degree of variability between temperature measurements, it would be advisable to keep the same inclination of the thermometer, e.g. at an angle of 45°, when taking the measurements, as temperatures are highly variable within the swath.
-   To increase the microporosity, or decrease the macroporosity, it is sometimes possible to regrind a certain amount of compost in order to reduce the particle size. The use of this substrate could result in a higher demand for nitrogen in the first weeks of production due to a lack of maturity of part of the compost.
-   If the compost is very compact (macroporosity too low), an aerator can be added on condition that it does not cause any change in the physico-chemical properties of the mature compost.
-   Chemical laboratory tests should be performed on the compost to better understand its properties and to facilitate the subsequent development of the fertilization program.

3.5- Linked to composting equipment :

-   The surface area of the composting slab is 600 m² is sufficient to install windrows with a volume of 400 ^{m3}.
-   Disinfection of the slab is required to avoid any form of contamination by undesirable organisms (nematodes and fungi).
-   The slope of the slab should be 1% to facilitate drainage of excess water.
-   The slab must be roofed to avoid excessive rainfall and sun exposure that induces desiccation of the surface portion of the swath.
-   In order to improve the shredding efficiency of the green material, it is necessary to use shredders with a minimum power of 20 HP.

3.6- Linked to the storage condition :

The storage area must be subject to conservation standards so that the compost does not lose its quality over time. The stowage of the compost must be :

-   Good aeration to avoid any possibility of rot or putrefaction or the release of frightening odors.
-   The storage slab should be raised off the ground to prevent rainwater from entering the compost to maintain optimal moisture.
-   Do not leave (compost) exposed to the sun to avoid drying. Thus keep a normalized moisture content.
-   It should be checked regularly and moistened from time to time, because compost is not a dead material like sand or gravel. It remains biologically active.

3.7- related to the use :

-   For sowing, it is advisable to sieve the compost. It will then be easier to use.

- Compost is not soil. You should avoid using pure compost but combine it with soil (1/3 compost, 1/3 soil, and 1/3 sand.
- Early spring is an excellent time to bring plenty of compost to the garden. Example: between rows of vegetables.
- In summer, mulching is the first use of compost, to be used before the heat to reduce water evaporation, limit soil compaction and promote the supply of organic matter.
- In autumn and winter, the compost is simply spread on the soil to better capture the sun's heat and it will also provide fertilization from early spring (2 to 8 kg/m2).

References bibliographies :

A. Tremier, A. De Guardia, and P. Mallard. 2007. Indicators of organic matter stabilization during composting and compost stability indicators: critical analysis and usage perspectives. French National Institute for Agriculture Food and Environment (INRAE). France: s.n., 2007. pp. 106-108.

ADEME. 2014. Compost. Environment and Energy Management Agency. France: s.n., 2014.

Alexis F. and Louis H.E., Lili M. and Caroline V. 2014. Technical guide for on-site composting in Institutions, Commerce and Industries. 2014.

Ammari Y., Lamhammedi M.S., Akrimi N., Zine El Abidine A. 2003. Composting of forest biomass and its use as a growth substrate for the production of seedlings in modern forest nurseries. INGREF. Tunisia: s.n., 2003. pp. 103-114.

Anonymous. 2015. Compost - The basics of on-farm composting. Agriculture, Aquaculture and Fisheries. New Brunswick (Canada): n.s., 2015.

—. 2016. Program for Sustainable Land Management and Adaptation to Climate Change in the Sahel (PRGDT).technical booklet for compost production. Ouagadougou-Burkina: s.n., 2016.

AZN. 2011. Composting methodology at the Ferme pilote de Guiè (FPG). Zoramb Naagtaaba inter-village association. Integrating environmental protection into Sahelian agriculture. Burkina-Faso: s.n., 2011.

B., Leclerc. 2001. Guide des matières organiques. s.l. eds ITAB Technical Guide, 2001.

B., Vanderhofstadt D. and Bastin. 1987. Composting project at Cap Serrat. Tunisian-Belgian cooperation. ODESYPANO. Sejnane.Tunisia: s.n., 1987. p. 45.

Bastin, D. Thesis . 195 p. 1987. End-of-study thesis. Fabrication et utilisation du compost de broussailles. Fabrication et utilisation du compost de broussailles. Experimental site of Cap Serrat, Tunisia. Provincial Institute of Higher Agricultural and Technical Education, Belgium. Tunisia: s.n., 1987. p. 195.

Bernal M.P., Paredes C., Sanchez-Monedero M.A., Cegarra J. Maturity and stability parameters of composts prepared with a wide range of organic wastes. Bioressource Technology. 1998. Maturity and stability parameters of composts prepared with a wide range of organic wastes. Bioressource Technology. pp. 91-99.

Brinton, W. F. and Evans E. 2000. Plant performance in relation to depletion, CO2-rate and volatile fatty acids in container media composts of varying maturity. 2000.

D., Jean. 1993. Methods for assessing compost maturity. Ecological agriculture projects. McGill University (Macdonald Campus). Quebec-Canada: n.d., 1993.

Dalzell, H. W., Biddlestone, A. J., Gray, K. R. et Thurairajan, K. 1988. Soil management: compost production and use in tropical and subtropical environments. Food an Agriculture Organization of the United Nations. Rome.Italie : s.n., 1988. p. 177.

Das, K. and H. M. Keener. 1997. Moisture effect on compaction and permeability in composts. 1997. pp. 275-281.

De Bertoldi M., Vallini G. et Pera A. 1983. he biology of composting: a review. Waste Management and Research. 1983. pp. 157-176.

Eggen, T. and O. Vethe. 2001. Stability indices for different composts. Compost Science et Utilization. 2001. pp. 19-26.

Eklind, Y. et Kirchmann H. 2000. Composting and storage of organic household waste with different litter amendments. I: carbon turnover. Bioressource Technology. 2000. pp. 115-124.

F., Canet R. and Pomares. 1995. Changes in physical, chemical and physico-chemical parameters during the composting of municipal solid wastes in two plants in valencia. Bioressource Technology. 1995. pp. 259-264.

F., Cédric. 2004. Stabilisation de la matière organique au cours du compostage de déchets urbains : influence de la nature des déchets et du procédé de compostage-Research for relevant indicators. Institut National Agronomique Paris-Grignon. France: s.n., 2004.

G., Schorth. 2003. Decomposition and nutrient supply from biomass. In G. Schorth & F.L. Sinclari, eds. Trees, crops and soil fertility: concepts and research methods. s.l. : CABI Publishing, 2003.

G.B., Willson. 1989. Combining raw materials for composting. s.l. : Bio-Cycle, 30, 1989. pp. 82-83.

Hirai, M. F., A. Katayama, et H. Kubota. 1986. Effect of compost maturity on plant growth. 1986. pp. 58- 61.

Hoitink H. J. A., Inbar Y. et Boehm M. J. 1991. tatus of compost-amended potting mixes naturally suppressive to soilborne diseases of floricultural crops. 1991. pp. 869-873.

ITAB. 2001. Guide to Organic Materials . 2001.

—. 2003. Use of compost in organic viticulture. 2003.

J.M., Paillat. 1997. Composting of organic materials of animal origin: environmental balance sheet. 1997.

K., Solbraa. 1986. Bark as growth medium. s.l. : Acta Horticulturae 178, 1986. pp. 129-135.

Kaiser, P. 1981. Microbiological analysis of composts. Report of the international symposium: Composts, humic and organic amendments. 1981. pp. 43-71.

Kapetanios, E. G, Loizidou M., and Valkanas G. 1993. Compost production from greek domùestic refuse. s.l. : Bioressource Technology, 1993. pp. 13-16.

Larsen, K. L. and McCartney D.M. 2000. Effect of C/N ratio on microbial activity and N retention in benchscale study using pulp and paper biosolids. s.l. : Compost Science & Utilization, 2000. pp. 147-159.

Leclerc, B. 2012. Composting: Principle. 2012.

Lemaire F., Dartigues A., Rivieres L.M. and Charpentier S. 1989. Culture in pots and containers. Agronomic Principles and Applications. INRA. Paris: s.n., 1989. p. 181.

Lemhamedi M.S., Fecteau B., Godin L. and Gingras C. 2006. Guide pratique de production en hors sol de plants forestiers, pastoraux et ornementaux en Tunisie. Direction générale des forêts. Tunisia: Pampev internationale ltée.Québec-Canada, 2006.

Mr., Mustin. 1987. Le compost. Management of organic matter. Paris: Editions François DUBUC, 1987. p. 954.

Mustin, M. 1987. Le compost . 1987.

Navarro, A. F., Cegarra J., Roig A., et Garcia D. 1993. Relationships between organic matter and carbon contents of organic wastes. s.l. : Bioressource Technology, 1993. pp. 203-207.

O., Verdonck. 1983. Reviewing and evaluation of new materials used as substrates. s.l. : Acta Horticulturae. 150, 1983. pp. 467-473.

Oueslati M.A., Ksontini M., Haddad M. and Charbonnel Y. 1995. Composting of Acacia cyanophylla branches and fresh sludge from wastewater treatment plants. s.l. : Rev. For. Fran. 5, 1995. pp. 523-529.

Palm C.A., Gachengo C.N., Delve R.J., Cadisch G. et Giller K.E. 2001. Organic inputs for soil fertility management in tropical agroecosystems: application of an organic resource database. s.l. : Ag. Ecosys. Env., 83, 2001. pp. 27-42.

R.T., Haug. 1980. Compost engineering. Priciples and practice. 1980.

R.V. Misra, R.N. Roy, and H. Hiraoka. 2005. Farm-level composting methods. United Nations Food and Agriculture Organization (FAO). 2005.

R.V., Misra. 2005. Farm-level composting methods. Land and Water Working Papers. Regional Office for Asia and the Pacific FAO. Bangkok: s.n., 2005.

R.W., Jeris J.S. et Regan. 1973. Controlling environmental parameters for optimum composting. Moisture, free air and recycle. s.l. : Compost Science.March-Avril, 1973. pp. 8-15.

Richard T.L., Hamelers H.V.M., Veeken A., et Silva T. 2002. Moisture relationships in composting processes. s.l. : Compost Science & Utilization, 2002. pp. 286-302.

Richard, T. 1996. The effect of lignin on biodegradability. 1996.

Roletto, E., Chiono R. et Barberis E. 1985. Investigation on humic matter from decomposing poplar bark. s.l. : Agricultural Wastes 12, 12, 1985. pp. 261-272.

Sadaka, S. et El.Taweel A. 2003. Effect of aeration and C:N ratio on household waste composting in Egypt. s.l. : ompost Science & Utilization, 11, 1, 2003. pp. 36-40.

Sanchez-Monedero, M. A., Roig A., Paredes C., et Bernal M.P. 2001. Nitrogen transformation during organic waste composting by the rugers system and its effects on pH, EC, and maturity of the composting mixtures. s.l. : Bioressource Technology, 78, 2001. pp. 301-308.

Sidhu, J., Gibbs R,A.,Ho G. E.et Unkovich I. 1999. Selection of Salmonella Typhimurium as an indicator for pathogen regrowth potential in composted biosolis. s.l. : Letters in Applied Microbiology, 29, 1999. pp. 303-307.

Stentiford, E. I. 1996. Diversity of composting systeme. In Science and Engineering of Composting. Blackie Academic and Professionnal. Bologne : s.n., 1996. p. 95.

T., Richard. 1996. The effect of lignin on biodegradability. 1996.

TRAME. 2008. Objective composting . 2008.

Chapter 3 :

Forest Reforestation

Table of Contents

List of tables

List of figures

1- Introduction :

The forest, in its tree-like entity, is the symbol of life for all living beings; it therefore provides shelter for certain species such as insects (tree trunk) and on the one hand, as a source of food for livestock and also provides **a** source of income for the benefit of local people through the law **of usage rights** (exploitation of dead wood lying on the ground and grazing). It makes vital contributions both to the populations and to the planet by purifying the air and preserving biodiversity, and offers solutions to cope with climate change.

However, today, forests are under great pressure due to human population growth, which is leading to their degradation in favour of unsustainable land use patterns.

Forest and land degradation is a serious problem, particularly in developing countries. In 2000, the total area of degraded forests and woodlands in 77 tropical countries was estimated at about 800 million hectares, including some 500 million hectares of primary and secondary forests. Forest degradation is one of the major sources of greenhouse gas (GHG) emissions, although its role has not been estimated on a global scale.

In order to mitigate the phenomenon of ecosystem degradation, reforestation **work is** the main means of **rehabilitating and restoring the forest** environment and making the forest a source of goods and services for the beneficiaries, be they human or animal, which can ultimately lead to a "non-degraded" forest over time.

2- Role of the forest :

The total economic value of the benefits provided by Tunisian forests has been estimated for 2012 at **208 million Tunisian Dinars** (TD), which represents **0.3% of GDP. This** corresponds to an economic value of about **176 TND/ha.**

2.1-   Economic role :

2.1.1- Source of supply of wood products :

Forests in general, and Tunisian forests in particular, are the only sources that provide wood products for various assortments, such as **timber** (for construction), wood for pallets or **pulpwood** (for the manufacture of particle board and paper), and **firewood** or charcoal for use by local populations.

2.1.2- Grazing :

The forests are also used as grazing **land and** thus contribute to the feeding of the livestock. **In 2004, the surface area of forest rangelands was estimated at 97,000 ha.**

2.1.3- Source of supply of NTFPs :

Tunisian forests also provide many Non-Timber Forest Products (NTFPs), the most important of which are cork, rosemary, myrtle, masticum, pine nut and Aleppo pine cones, mushrooms, snails and honey. These NTFPs thus provide revenue to the Tunisian State. Nevertheless, the economic importance of these NTFP products is still undervalued.

2.1.4- Hunting :

Hunting activities carried out in Tunisian forests can also generate income linked to the marketing by the owner of the game killed or hunting rights.

2.2-   Ecological role :

Forest ecosystems play an essential role in **protecting soils against erosion** (especially water erosion in mountainous areas), **soil fertility** through the decomposition of plant organic matter (leaves and twigs), making the soil more stable and cohesive and reducing the risk of landslides; thus in the **preservation of water resources** (combating silting of dams).

Forests also have a substantial role to play in the **fight against climate change,** through the sequestration of carbon in vegetation and soil, and in **adaptation to climate change, in the** face of the significant rise in temperature and decrease in rainfall by 2020 and 2050 predicted by climate change projection models.

2.3-   Cultural role :

Forests play an important **recreational role**, which is reflected in the generation of income linked to the commercialisation of forest access rights and the indirect economy linked to the recreational function of the forest and tourism (payment of guides' and guards' salaries, money spent by visitors injected into the local economy, etc.).

Forests provide services related to the **conservation of** fauna and flora **biodiversity.** Tunisia's terrestrial flora includes 2,200 species. The Tunisian terrestrial fauna includes more than 500 species of vertebrates, including nearly 400 species of birds, nearly 80 species of mammals and more than 60 species of reptiles. Insects are also represented by about 670 species. Tunisian forests thus play an essential role in the preservation of this biological heritage, mostly sheltered by forests and rangelands.

Table 2: Estimated annual benefits and costs of forest degradation in Tunisia (in 2012)

Benefits and costs	Total value (in millions of DT)	Value per hectare (in DT/ha)
Benefits		
Procurement Services	128,9	109
Regulatory Services	74,9	63
Cultural Services	4,4	4
Total benefits	**208,2**	**176**

3- Current situation of the Tunisian sylvopastoral resource :

The surface area by type of land use from the second national forestry and pastoral inventory "IFPN" shows that the area of forest land was thus estimated at **1,141,628 ha (between 1998 and 2003),** whereas **purely forest formations** (including scrublands and wooded scrublands) would in reality only cover **673,193 ha (or 4.1 % of the national territory).**

Table 3: Area distribution by land use according to the results of the second FNFPI

Land use	Surface area (ha)	Proportion (%)
Forest areas	**673 193**	**4,1**
Arboretum	480	0,0
Boqueteau	1 908	0,0
Deciduous forest	140 209	0,9
Softwood forest	374 862	2,3
Mixed forest	26 633	0,2
Wooded scrubland	34 383	0,2
Young settlement	66 901	0,4
Scrubland with trees	13 007	0,1
Plantations of banks, dunes	14 810	0,1
Other forestry training	**464 436**	**2,8**
Windbreaks, hedges	1 831	0,0
Forest clearings	1 207	0,0
Garrigue without trees	243 892	1,5
Uncultivated land	131 855	0,8
Forestry infrastructures	8 610	0,1
Scrubland without trees	70 178	0,4
Alignment planting	6 863	0,0
Other land	**15 258 372**	**93,1**
Built land	184 694	1,1
Cultures	4 503 112	27,5
Desert	4 555 957	27,8
Course + Mosaics	5 213 830	31,8
Water and wetlands	800 778	4,9
Grand Total	**16 396 000**	**100,00 %**

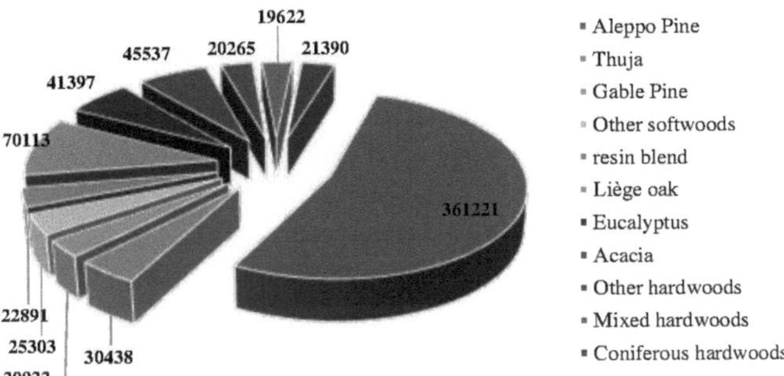

Figure 25: Distribution of forest areas according to their specific composition (DGF; 2010)

4- Terms of Reference :

4.1- Deforestation :

4.1.1- Definition :

i. Deforestation or the reduction of forest plant cover; is the action of clearing forest areas without concern for the renewal of these forests. Deforestation leads to a huge loss of forest area in the world.

ii. Deforestation is the set of practices and processes leading to the irreversible loss of forested areas to non-forest activities, generally agriculture and urbanisation. According to the FAO, the term applies when forest cover reaches a density of less than 10% per hectare.

iii. Deforestation is the phenomenon of reduction of forest areas. We speak of deforestation when areas of forest are definitively lost or at least lost in the long term.

4.1.2- Causes of deforestation :

In the world, the loss of forest areas, deforestation is caused by multiple factors, such as :

➤ **Natural** :
  - Forest Fires ;
  - Diseases : Biscogniauxia mediterranea, Mycosphaerella pini..., etc.
  - Insect attacks : Platypus cylindrus, disparate Bombyx, etc.

➤ **Anthropic**
  - Subsistence agriculture would account for 30 to 35% of global deforestation;
  - Commercial or industrial agriculture (field crops and livestock) would account for 45 to 50% of deforestation;
  - Livestock is the cause of about 14% of deforestation worldwide;
  - The construction of infrastructures would represent about 8% of deforestation;
  - Mining activities would be responsible for approximately 6% of the forest loss;
  - Urbanization around 5%.

4.1.3- Consequences of deforestation :

4.1.3.1- On biodiversity :

Forests are considered to be one of the richest hubs of biodiversity in the world because they are home to living beings: mammals, birds, insects, plants and sometimes rare species.

Indeed, human activities threaten the existence of these species and this can have important consequences on natural balances. Thus, in some regions of Africa, great apes such as the silver gorillas are endangered because of the progressive reduction of their natural habitat, particularly due to deforestation.

4.1.3.2- On floors :

The presence of a forest makes the soil richer in organic matter, more consistent and more resistant to weathering or erosion. As a result, when a forest area is destroyed, the soil has become very fragile and little by little the ecosystem will be more vulnerable to natural disasters such as landslides or floods.

4.1.3.3- On the climate :

Indeed, trees store CO_2 throughout their life. By destroying these trees, we therefore reduce the capacity of the global ecosystem to store CO_2. Fewer trees means less CO_2 absorbed and therefore more greenhouse effect. As a result, it is estimated that deforestation is responsible for the equivalent of 11.3% of global anthropogenic CO_2 emissions, making it one of the largest contributors to global warming on a par with road transport and the energy consumption of buildings.

4.2- Reforestation :

It is the restoration of forest formations after a forest disturbance of human or natural origin.

4.3- Afforestation :

Afforestation or afforestation is the opposite of deforestation. It is the establishment of forest plantations on land previously **not classified as forest, through** silvicultural measures, with forest cover **above the 10% threshold.** The term implies the transition from a state of non-forest to one of forest proper.

4.4- Deforestation :

Deforestation; means the cutting or felling of all trees in a place or forest without a program of restocking or regeneration.

4.5- Reforestation :

It is the restitution of a forest environment after a clear cut or final cut.

5- Reforestation :

5.1- Definition :

It is an operation that consists of restoring or creating wooded areas that have been removed by clear cutting, or destroyed following a disaster. Reforestation can be natural or artificial.

5.2- Type of reforestation :

5.2.1- Natural Reforestation :

Is a spontaneous regeneration, **without human intervention**. It is done by the dissemination of seeds via **wind, water or animals**.

5.2.2- Artificial reforestation :

Replanting of shrubs to restore the density of the plant cover. This type of reforestation is carried out by man to :

- ✓ Restore lumber stocks or for
- ✓ Stabilize soils eroded by agricultural and livestock activities; or by the
- ✓ Deforestation, particularly due to the removal of a large number of trees for the
- ✓ Consumption (coal, firewood, etc.).

5.3- Reforestation objectives :

5.3.1- Socio-economic :
- ✓ Aim to meet the growing need for firewood and service wood ;
- ✓ Improvement of fodder production for livestock when it is lacking;
- ✓ Involvement of the population in the reforestation program and consideration of their needs;
- ✓ Reduces the unemployment rate even if occasionally ;
- ✓ Commitment of the populations of the rural community to the protection of the forest, on the one hand through their contribution to the reforestation project and, on the other hand, a feeling of responsibility that the forest is the only source of life for them and their livestock and the right to use certain forest products (reduction of the deficit in wood for services and energy) ;

5.3.2- Environmental :
- ✓ Recovery of degraded forest land ;
- ✓ Rehabilitation of degraded forest ecosystems ;
- ✓ Restore the ecological balance again
- ✓ Decrease in the time needed to develop the new forest

5.3.3- Cultural :
- ✓ To master the techniques of growing plants in pine forest;
- ✓ Highlighting on all in situ follow-up work from the time of transplantation.

5.4- Limitations of reforestation :
- ➢ **Morphological and physiological properties of the plant at the exit of the nursery :**
  - Cultivation techniques ;
  - Storage conditions ;
  - Conservation of plants.
- ➢ **Effect of environmental factors on these characteristics after planting :**
  - Adaptation of the gasoline to the station ;
  - Good preparation of the ground ;
  - Wise planting period.
- ➢ **Accompanying work :**
  - Irrigation,
  - Fertilization ;
  - Defense ;
  - Sanitary treatment.
- ➢ **External Factors :**
  - Transplant condition ;
  - Climate ;
  - Parasitic attacks (insects and fungi).

5.5- Planning the reforestation program :

5.5.1- Preparation of the plants for the planting site :

We start by sorting the plants of plantable size. This sorting depends on standards; such as size, resistance to climatic hazards: wind, drought ...). The main objectives of a sorting system for planting material are the following:

- ✓ Eliminate, damaged or sick plants
- ✓ Eliminate the plants that do not reach the minimum standards of size and root development;
- ✓ Classify plants that exceed the minimum standards into two or more quality categories.

5.5.2- Site preparation :

Before starting the work, a schedule of all the tasks to be performed is first established. The organization of the work site must be done according to time and space, while taking into consideration the budget reserved for the tasks to be performed.

- ❖ **In time:** The various stages of site preparation must begin and end in time.
- ❖ **In space:** All tillage operations must touch the entire area to be replanted.

The duration of this phase depends on several factors, as follows:

- Availability of workers ;
- The quality of the workers ;
- Nature of terrain (soil texture, topography) ;
- Availability of materials ;
- Climatic conditions...

5.5.3- Site preparation :

Site preparation is considered to be the cradle of the seed so that it can germinate, resulting in a stand that meets reforestation objectives. This stage has a large scope (extension) in view of their importance to seeds.

Moreover; the preparation of the ground is an indisputable step that should never be neglected before each reforestation program. Not only does it increase the chances of survival of the young plants, but it also facilitates the work of reforestation.

Indeed, if one can have desirable results, it is essential to mobilize people who are specialists in soil preparation to lead the land suitable for the resumption of planting and on the other hand to ensure proper growth.

5.5.3.1- Tillage objectives :
- ✓ Increase the water retention capacity ;
- ✓ Ensuring deep root anchoring ;
- ✓ Improving the chemical composition of soil ;
- ✓ Control runoff and erosion ;

5.5.3.2- Soil working techniques :
5.5.3.2.1-   Ground clearance :
5.5.3.2.1.1- Characteristics and objectives :

Debris removal consists of piling up dead branches and other residues from the cut in strips called "strips". Clearing allows :

- Facilitates the planting of plants;
- Produce a favourable environment for plant growth ;

5.5.3.2.1.2- Release technique :

This type of preparation can be done with a comb mounted on a tractor. A straight or "V" blade mounted on a tractor can be used to stack the cuttings.

5.5.3.2.1.3- Precautions when clearing :

When heaping waste, it is important not to scrape the soil down to the mineral soil. The mineral soil is located under the peaty layer of little or no compound debris. It is a layer that retains very little water. Thus, if the mineral soil is exposed, surface water will not be absorbed at depth, hence runoff will be produced on the ground.

Figure 26: Comb mounted on a tractor

5.5.3.2.2- Clearing :

5.5.3.2.2.1- Characteristics and objectives :

Brushing is the removal of all undesirable vegetation from the reforestation surface that can induce competition with the desired species (nutrients, water, light), which then creates growth risks.

5.5.3.2.2.2- Brushcutting technique :

The removal of unwanted plants can be done manually or mechanically. The choice of technique depends on the area treated and the type of vegetation to be removed.

The manual release can be done with the swedish knife, machete or pruning shears, the axe is strongly discouraged due to the risks of accidents that may occur. The removal of plants must be with their root systems (uprooting), so that their growth is not fast. This technique is recommended, and less expensive if the surface is small. In a reforestation project this method will be very expensive, disadvantageous; because we are talking about tens of hectares to treat, in this case it is essential to choose the mechanical method, using the brushcutter and shredder.

The field treatment is done between May and the beginning of September to eliminate the risk of stem re-growth from the previous day.

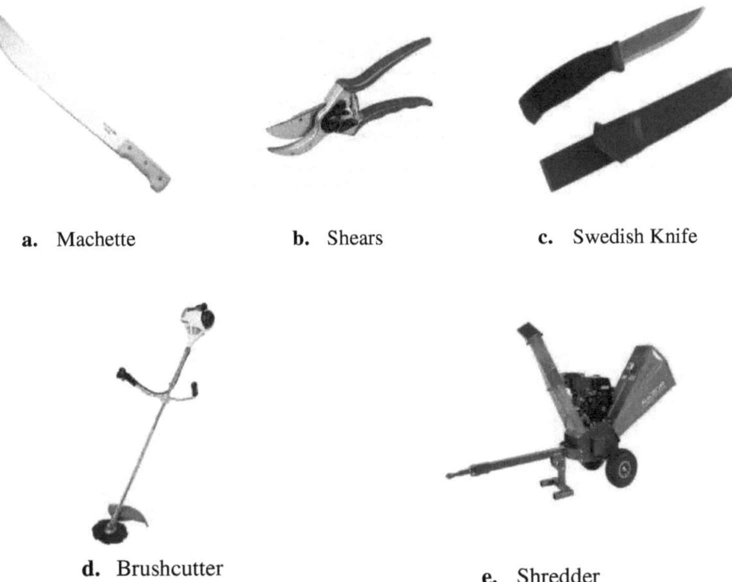

Figure 27: Brushcutting equipment

5.5.3.2.3- Floor stripping :
5.5.3.2.3.1- Features :

Stripping is a method of removing the top layer of the soil, which is mainly composed of plant matter (20 to 40 cm), without damaging the substrate, which is the inner layer of the soil.

5.5.3.2.3.2- Pickling objectives :

The objectives of the stripping work are :

- ✓ Lowering the ground level, which results in a relative increase in the height of the water table. Depending on the height and thickness of the layer removed, the hydromorphy of the soil is less and less deep and thus allows a better vegetation on the surface.
- ✓ To improve environmental conditions, thus contributing to the diversification of habitats and species.
- ✓ To expose the ground to allow the relocation of pioneer stages of vegetation.

5.5.3.2.3.3- Main steps of pickling :

There are several steps to follow in the stripping process:

- ✓ The location and delimitation of the area to be stripped are important in the sense that stripping requires preliminary studies at the level of soil diagnosis in terms of pedology, hydrology and ecology.
- ✓ Stripping can be carried out manually or mechanically depending on the sensitivity and constraints of the terrain, the means available in terms of manpower and budget, as well as the size of the work on the surface and the volume of material to be stripped.
- ✓ The collection and export of the collected material is done with a dumpster or dumper and exported off-site. This export can be done manually, semi-mechanized or mechanized.
- ✓ Post-construction follow-up consists of checking whether the stripping has been successful or still requires adjustment.

5.5.3.2.3.4- Pickling techniques :

There are three main pickling techniques, which are mainly dependent on the materials used

- ✓ **Pickling with a mini-excavator :** is mainly adapted to sites with low bearing capacity;
- ✓ **Backhoe stripping:** is possible for relatively load-bearing soils;
- ✓ **Excavator stripping:** allows for greater site output.

| **a.** Mini- | **b.** Backhoe | **c.** Backhoe |

Figure 28: Soil stripping equipment

5.5.3.2.4- **Basement :**
5.5.3.2.4.1- **Features :**

Subsoiling is an agricultural technique used to restore soil permeability by improving the natural drainage and horizontal capillary circulation of water on ploughed soil, and to regenerate the structure of the soil horizons under the plough bed. This operation concerns only the layer of soil above the ploughing bed, which varies in depth from 50 to 85 cm.

5.5.3.2.4.2- **Subsoiling objectives :**

The purpose of subsoiling is to regenerate the structure of the soil that may have been deteriorated at depth, in particular due to :

- ✓ Improve deep root growth ;
- ✓ Promote the drainage of excess water.

5.5.3.2.4.3- Technique of subsoiling :

Subsoiling must be done on dry soil or on a non-plastic material. If you have to work at a depth of **30-35 cm**, you will talk about **decompaction,** which is usually done with a chisel. **At 50 cm and more up to 85 cm** deep, we speak in this case to subsoiling carried out by the subsoiler. Although the subsoiling it consists in the opening of planting holes, it can use the chemical method (ammonium nitrate). A large volume of soil is thus disturbed, which favours the development of the root system of the plants.

a. Subsoiler   b. Chisel

Figure 29: Undergrounding equipment

Table 4: Summary of tillage techniques

Ground Clearance				
Features	**Objectives**	**Techniques**	**Materials**	**Season**
- Consists of piling up dead branches, old logs and other residues from the cut.	- Facilitates the planting of plants; - Produce a favourable environment for plant growth ;	- Stack cutting debris in "swathes".	- Comb mounted on a skidder or crawler tractor. - A straight or "V" blade mounted on a crawler tractor	Dry
Clearing				
Features	**Objectives**	**Techniques**	**Materials**	**Season**
- Removal of all undesirable vegetation from the reforestation field.	- Avoid competition between undesirable species (grasses) and plants to be reforested. - Ensuring good and rapid growth	- Removal of grasses by these root systems	- Swedish knife, machete, pruning shears **(Manual).** - Chipper and shredder **(Mechanical).**	May - September

Floor stripping				
Features	**Objectives**	**Techniques**	**Materials**	**Season**
- Removal of the surface layer of soil Remove all traces of vegetable waste	- Facilitates the installation of pioneer plants. - Improvement of environmental conditions. - Lowering the ground level ⇨ Relative increase in the water table.	–	- Mini-excavator - Backhoe loader - Backhoe.	Dry

Under soil relief				
Features	**Objectives**	**Techniques**	**Materials**	**Season**
Agricultural technique is used to restore soil permeability and drainage.	- Improves deep root growth. - Promotes drainage. Regeneration of the soil structure	Carry out on dry earth or non-plastic materials. - **Decompacting: (30 to 35 cm).** - **Under relief: (50 to 85 cm).**	- **Chisel : Decompacting** - **Subsoiler : Under relief** - **Ammonium nitrate**	Dry

5.5.2- Elaboration of the holes :
5.5.2.1- Characteristics and objectives :

After having a land suitable for reforestation, we move on to the phase of burying the holes, this is the last location for the plants. The holes must be less than 30cm deep, so that the young seedlings are fixed, and the roots are well anchored in the soil.

5.5.2.2- Technique of elaboration of the holes :

There are several techniques for burying holes for reforestation. If the manual method is used, provided the surface area is reduced, the **drilling shovel** can be used.

- Easy to handle,
- Less expensive,
- Less expensive ;
- Risk of accidents almost zero.

To speed up the work, it is preferable to use the **electric auger or drill**.

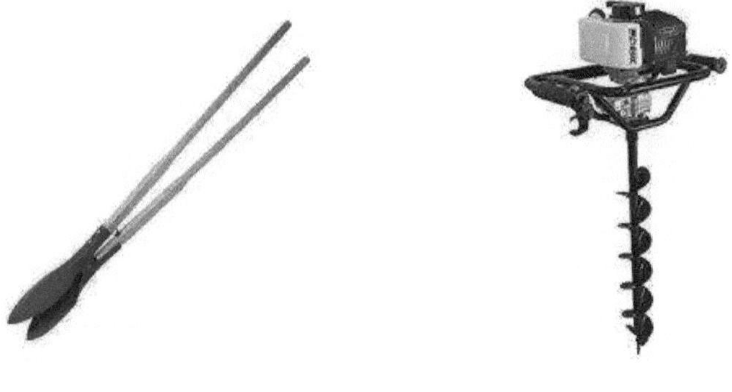

a. Drilling shovel   **b.** Electric auger

Figure 30: Materials for the elaboration of planting holes

5.5.2.3- Hole spacing :

The spacing between seedlings is varied according to several criteria, for an objective threshold the success of the reforestation program and possibly avoid or reduce the replenishment as much as possible. Among these criteria that must be taken into account :

- **The species:** water requirements, response to water stress...
- **Soil:** texture, water retention capacity
- **Climate:** annual rainfall...
- **Topographic:** slope, exposure...

Several types of spacings were used by the project (3mx2m, 3mx3m, 4mx4m, etc.). The number of seedlings planted per hectare is 1666, 1111 and 625 respectively.

5.5.3- Transplantation of plants :
5.5.3.1- Characteristic :

It is the transport of young seedlings from the nursery to the reforestation site.

5.5.3.2- Precautions for use :

At the time of transplantation, it is necessary to :

- ✓ Avoid mishandling when loading and unloading vehicles.

- ✓ The plants must be protected during the transport, because the draught can dry them out.

- ✓ It is important to tighten the containers to prevent them from moving. Special shelves can be added to the vehicle platform to place the pots or boxes (each layer of boxes is placed on a shelf, with the shelves 50 centimetres above each other).

- ✓ The plants should be transported during the planting season when the weather is cool, cloudy or even rainy, to prevent drying out during transport.

- ✓ The best successes in reforestation were obtained with young seedlings (less than one year old from the nursery).

- ✓ Planting was mainly done during the rainy season. It started in mid-October (big rainy season) and continued until the end of February (small rainy season).

- ✓ When placing the plants in the field, it is preferable to section the bottom of the root ball carefully to avoid any risk of removing the hairs of the plant which can cause growth trauma. One can also break the root ball just at the time of the plantation and unroll the root system, the roots being exposed to the air only during a few seconds.

5.5.4- Accompaniment of the reforestation program :
5.5.4.1- Defense :

The protection of seedlings is a precaution that must be taken into consideration before, during and after reforestation, as long as the young plants have not yet reached the survival stage, or the young subjects have become resistant to the damage of predators.

Circling the entire reforested area is fundamentally required to protect the young plants from predators (large and small game), it is done by barbed iron wire with a height of 2.50 meters, supported by iron posts well anchored in the ground to guarantee the future forest and the objectives set reached its terms.

It is necessary to protect the most palatable species (especially Pines and Oaks) with individual wire mesh sleeves about 20 centimeters in diameter and 60 in height, securely fastened with two diametrically opposed stakes.

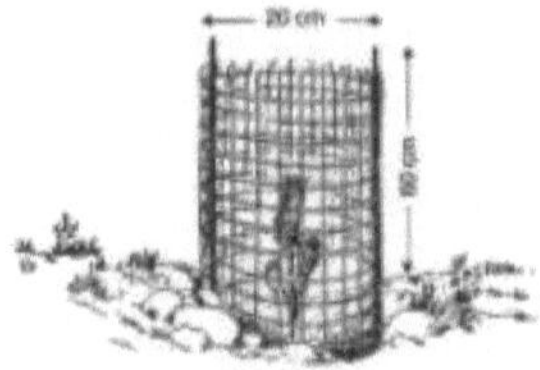

Figure 31: Individual sleeve

5.5.4.2- Interviews :
5.5.4.2.1-   Irrigation :

In reforestation, irrigation is essential, especially in the heat wave period when young plants are still weak and unable to tolerate the water deficit.

The supply of irrigation water to the reforestation site is relatively complicated for several reasons, either economic, such as equipment costs (water pump, irrigation pipes), or water-related considerations (quality, volume, flow rate, etc.).

5.5.4.2.2-　Fertilization :

The quality of the soil is a limiting factor that guides the growth of the plant, if the soil is well fertile we will have plants with a rapid growth, phenotypically vigorous and meet the objectives of reforestation.

The deficiency in mineral elements is expressed by the symptoms that appear on the plant (yellowing, leaf wilt, slow growth, etc.). If the damage is individualized (a few plants), we target the plants in question with the treatment compatible with their situation, if the damage is extensive we put the nutrients in the irrigation water.

5.5.4.2.3-　Sanitary treatment :

After establishment, the development of young trees is subject to various risk factors, the most dangerous of which are competitive vegetation and pests.

5.5.4.2.3.1- Against competitive plants :
5.5.4.2.3.1.1-　Hoeing :

It consists of uprooting grass with shallow tillage. The hoeing is mechanical by the continuous action of the tractor between the lines of plants, or manually by hoeing in the immediate vicinity of the plants.

5.5.4.2.3.1.2-　Phytocide :

Consists of spreading on the surface to be treated a preparation containing a certain percentage of active product. This product, necessarily selective so as not to damage the species introduced, will differ according to the species (more or less sensitive to this or that speciality) and according to the stage of development of the weeds, i.e. according to whether it is treated before or after their germination.

It consists in coating the surface to be treated with a preparation containing a certain percentage of active product. This product is necessarily selective so as not to damage (harm) the introduced species and according to the stage of development of the weeds.

5.5.4.2.3.2- Against parasitic attacks :

The young plants are not in a closed vase, i.e. they are always confronted with the attacks of either entomological or pathological pests. They must be treated as soon as they appear and the epidemic remains isolated (a few feet).

5.5.4.3- Regarnis :

The success or failure of the reforestation project has already been discussed before the start of the market. The success of reforestation is discussed if the rate of well-anchored seedlings developed after one year from the date of reforestation exceeds 85%. On the other hand, if the failure rate of the seedlings in the same period is 40% or more, i.e. the success rate does not exceed 60%, then the market will be terminated.

The operation of replenishment, it gathers the transplanting in nursery. The condition of replenishment, if the success rate of the plants in the period that is already mentioned below is between 60 and 85%. The replenishment is a substitution of dead plants by other living plants, thus increasing the success rate more than 85%.

5.5.4.4- Validation :

It can be said that the reforestation project is validated if all the plants that make up the future forest have a normalized growth, free from parasitic attacks. The approval of the project is not counted until after 5 years from the establishment or the young plants have reached the stage of thicket and gaulis.

References Bibliographies

Anonymous; 2016: Forestry Investment Program in Tunisia".

Christine A., Gérard F. and Jean G.; 1991: Production de plants forestiers: Guide technique du forestier méditerranéen français.

FAO; 1992: Forestry in drylands. A guide for field technicians.

FAO; 1998: Community Reforestation Project in the Precoba-Senegal Groundnut Basin. Food and Agriculture Organization of the United Nations-Rome-Italy.

Nicolas S.O.; 2003: The environmental and socio-economic benefits of a planted forest - the case of the 8,000-hectare reforestation project on the Batéké-Kinshasa plateau. Democratic Republic of Congo.

Robert M.; 1992: Reforestation techniques. Technical guide for French Mediterranean foresters. Center National du Machinisme Agricole du Génie Rural des Eaux et Forêts-CEMAGREF.

More Books!

I want morebooks!

Buy your books fast and straightforward online - at one of world's fastest growing online book stores! Environmentally sound due to Print-on-Demand technologies.

Buy your books online at
www.morebooks.shop

Kaufen Sie Ihre Bücher schnell und unkompliziert online – auf einer der am schnellsten wachsenden Buchhandelsplattformen weltweit! Dank Print-On-Demand umwelt- und ressourcenschonend produziert.

Bücher schneller online kaufen
www.morebooks.shop

KS OmniScriptum Publishing
Brivibas gatve 197
LV-1039 Riga, Latvia
Telefax: +371 686 204 55

info@omniscriptum.com
www.omniscriptum.com

Printed by Books on Demand GmbH, Norderstedt / Germany